Macmillan Computer Science Series

Consulting Editor

Professor F. H. Sumner, University of Ma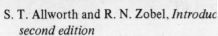

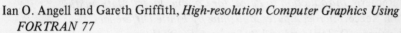

S. T. Allworth and R. N. Zobel, *Introduc*
   second edition

Ian O. Angell and Gareth Griffith, *High-resolution Computer Graphics Using FORTRAN 77*

Ian O. Angell and Gareth Griffith, *High-resolution Computer Graphics Using Pascal*

M. A. Azmoodeh, *Abstract Data Types and Algorithms*

Philip Barker, *Author Languages for CAL*

A. N. Barrett and A. L. Mackay, *Spatial Structure and the Microcomputer*

R. E. Berry and B. A. E. Meekings, *A Book on C*

G. M. Birtwistle, *Discrete Event Modelling on Simula*

T. B. Boffey, *Graph Theory in Operations Research*

Richard Bornat, *Understanding and Writing Compilers*

Linda E. M. Brackenbury, *Design of VLSI Systems – A Practical Introduction*

J. K. Buckle, *Software Configuration Management*

W. D. Burnham and A. R. Hall, *Prolog Programming and Applications*

J. C. Cluley, *Interfacing to Microprocessors*

Robert Cole, *Computer Communications, second edition*

Derek Coleman, *A Structured Programming Approach to Data*

Andrew J. T. Colin, *Fundamentals of Computer Science*

Andrew J. T. Colin, *Programming and Problem-solving in Algol 68*

S. M. Deen, *Fundamentals of Data Base Systems*

S. M. Deen, *Principles and Practice of Database Systems*

Tim Denvir, *Introduction to Discrete Mathematics for Software Engineering*

P. M. Dew and K. R. James, *Introduction to Numerical Computation in Pascal*

M. R. M. Dunsmuir and G. J. Davies, *Programming the UNIX System*

K. C. E. Gee, *Introduction to Local Area Computer Networks*

J. B. Gosling, *Design of Arithmetic Units for Digital Computers*

Roger Hutty, *Fortran for Students*

Roger Hutty, *Z80 Assembly Language Programming for Students*

Roland N. Ibbett, *The Architecture of High Performance Computers*

Patrick Jaulent, *The 68000 – Hardware and Software*

J. M. King and J. P. Pardoe, *Program Design Using JSP – A Practical Introduction*

H. Kopetz, *Software Reliability*

E. V. Krishnamurthy, *Introductory Theory of Computer Science*

V. P. Lane, *Security of Computer Based Information Systems*

Graham Lee, *From Hardware to Software – an introduction to computers*

A. M. Lister, *Fundamentals of Operating Systems, third edition*

G. P. McKeown and V. J. Rayward-Smith, *Mathematics for Computing*

(continued overleaf)

Introduction to Pascal for Computational Mathematics

E. J. Redfern

Department of Statistics
Leeds University

**MACMILLAN
EDUCATION**

First published 1987

Published by
MACMILLAN EDUCATION LTD
Houndmills, Basingstoke, Hampshire RG21 2XS
and London
Companies and representatives
throughout the world

Typeset by TecSet Ltd, Wallington, Surrey
Printed in Hong Kong

British Library Cataloguing in Publication Data
Redfern, E. J.
   Introduction to Pascal for computational
   mathematics.—(Macmillan computer science
   series).
   1. Mathematics—Data processing   2. PASCAL
   (Computer program language)
   I. Title
   510'.28'55133   QA76.95

ISBN 0-333-44430-2
ISBN 0-333-44431-0 Pbk

Contents

Preface

Pascal is a rich and versatile computer language, designed to allow construction of programs that closely relate to the algorithm proposed as the solution to a particular problem. Well-written Pascal programs, therefore, are easy to read, understand and modify.

It is this feature of the language which makes it attractive as a first programming language. It encourages the programmer to carefully break problems into component parts which can then be translated into the programming language. As a result, it is best learnt alongside the process of learning how to solve problems numerically since the features of a problem's solution often have corresponding constructions in the language. Conversely, the problems supply suitable material for purposeful practice in using the computer.

This book has evolved out of a course on programming for first-year mathematicians at Leeds University. As well as teaching programming there was an additional requirement that the students be introduced to numerical problems related to their other courses. This performs the dual role of motivating those students who have not realised how important it is for the modern mathematician to be able to program, while allowing them to do numerical exercises related to material covered in their other courses.

The course, and subsequently the book, therefore were designed around the need to program numerical problems. The development of programming knowledge is dictated by the problems and no feature of the language, with the possible exceptions of procedures and functions, is introduced until it is required by the areas of application.

Thus, as well as introducing the Pascal language, this book is intended to serve as an introduction to solving numerical problems that arise in a first-year university Mathematics course. These problems are presented from a first principle point of view, with detailed proofs and justifications usually omitted.

The first part of the book (chapters 2–13) covers the material that I have at some stage included in the twelve lecture course given to our new Mathematics students at Leeds. It should however be stressed that not all the application areas have been covered in a single run of the course. This section covers the ideas of numerical methods such as iteration, integration, first-order differential equations and solving simultaneous equations. It is based solely on the use of INTEGER, REAL and BOOLEAN scalar types and ARRAYS, together with the constructions which allow looping and choice of alternative paths through a program.

The second part of the book (chapters 14-20) explores briefly the special features of Pascal such as RECORDs, POINTERs and SETs, again relating them to problems of interest to mathematicians. Thus complex numbers and Fourier transforms illustrate the use of RECORDs, and polynomial manipulation and queues are used to motivate and illustrate POINTERs. Finally, SETs and POINTERs are applied to examples involving trees and graphs.

Exercises are provided at the end of most chapters. Most of these are designed to give practice on the ideas developed in the preceding chapter, while others are intended to explore mathematical ideas numerically.

I would like to acknowledge the many helpful comments, suggestions and corrections made by colleagues and students. In particular, I would like to thank Patrick Constable, Geoffrey James, Allan Watkins, David Knapp, David Salinger, Tim Pullan and Stephen Mobbs who have read through various parts of the manuscript at some stage in its development. I would also like to express my appreciation of three independent reviewers for their criticisms, corrections and suggestions which I have found very useful. Naturally, any errors that remain are my responsibility.

Finally I would like to thank my wife Jane for her help and encouragement while I was writing the book.

1 Introduction

Learning to use a computer to solve problems involves both learning a programming language and learning how to write programs. Learning any language, spoken or computer, provides no motivation unless we are going to use it. This book is designed with both these objectives in mind. As well as introducing the reader to the Pascal programming language, we consider how to use that knowledge in solving problems that require the use of a computer. The text has evolved out of an introductory course on programming for mathematicians and hence the problems considered are related to the work that mathematics students are likely to cover during their first year in higher education.

When introducing a programming language the emphasis is best placed on quickly achieving sufficient programming skills to enable us to start using a computer without the need to assimilate a lot of theory about the language. In this text, therefore, we let the mathematical problems in which we are interested dictate how much of the language we need to have learnt at each stage. Thus, as each element of the language is introduced it is used to solve a variety of interesting problems, so ensuring that the reader can gain proficiency in its use.

Before we begin to introduce the language, however, we need some basic terminology related to using a computer and a programming language. A user of a modern computer usually works at a keyboard linked to a monitor which serves as a screen to display what is entered on that keyboard. There is also usually some form of storage device available on which to store data, results and programs. If we are working on a mainframe computer the storage device and the actual computer are remote from the user, while in the case of a microcomputer, they are part of the equipment in front of us. This difference, however, usually causes no fundamental change in the way that we use high-level programming languages such as Pascal.

Programs are entered on to a computer and altered using an editor. The text of the program can be stored in a file on the storage device, preserving it for future use. These features are dependent on the machine that you are using and hence are not covered in more detail in this text.

Once a program has been created as text on the computer, its text must be converted into a code that the computer understands. This is done by using a compiler which is a program available on our computer. As well as performing the conversion, it also checks the syntax of the program to see that it is consistent with the rules of the language. Pascal is a language designed so that

the compilation is robust and any errors in the syntax of the program are
detected at this stage.

A program is used to tell a computer how to perform calculations and
manipulate data. Programming languages like Pascal have been designed to
make this task easy. Pascal is also designed so that the resulting program is easy
to read and bears a close relationship to our proposed mathematical solution to
the problem. More specifically, a program consists of a set of statements which
are obeyed in order. Most programs can usually be described by three main
steps:

1. data entry,
2. calculation, and
3. display of results.

Each of these steps can be planned and worked out separately.

This structure is illustrated in our first program in chapter 2 which we use as
a basis for introducing the fundamental features of a Pascal program. We also
describe the programming structures that allow us to organise the entry and
display of data. It is necessary to deal with these problems at an early stage.
When we begin writing programs, however, we usually sort the calculations out
first. Only when we have the program working should we spend time on improv-
ing the way in which the program displays information.

The remainder of chapter 2 is spent in describing ways of carrying out simple
calculations, while in chapter 3 we consider the parts of the language that give
us ways of repeating groups of operations and choosing between alternatives.

Once we have introduced these relatively simple elements of the language we
are able, in chapter 4, to apply them to handling sequences and series on a computer. The methods described here are forerunners to those in later chapters,
which deal with solving non-linear algebraic equations and differential equations.
As well as discussing the programming techniques, we also begin to describe the
ways in which we can use these results and the interpretation that we can place
on them.

As well as learning how to carry out calculations, we also need to consider
how they are performed on a computer and the errors that can occur if we are
not careful. The main point to remember when using a computer is that any
real number is only held to so many significant figures. This results in rounding
errors, one consequence of which is that any arithmetic carried out on a computer does not produce results that exactly match those we would get performing the calculations on paper. In many instances this is not a serious problem,
but the sooner we are aware of it the better. Thus, in chapter 5, we digress a
little and describe how calculations are performed on a computer and the
possible errors that can result.

The next two chapters return to introducing features of the Pascal language.
As soon as we start writing programs we start generating results which, in many
instances, can be extensive. We also need to be able to enter large amounts of

data into programs, which can be tiresome if they are entered at the keyboard while the program is running. (We are also likely to make errors that can only be corrected by typing all the data in again.) We therefore introduce the concept of a data file which can be used both for entering data into the program and for storing results that are generated by the program. Once we have results in a file we can study them on the screen or print them out when required. It is very easy to do this within a Pascal program but the way in which an external file is linked to the program depends on the computer that we are using and again cannot therefore be described within this text. Pascal itself has more extensive file-handling capabilities, which we are not going to describe since they are not required for the problems that we are considering.

As well as looking at ways of preserving results we also consider ways of ensuring they are neatly laid out. One common form of display is the results table, and if we can generate this neatly within our program it makes the later presentation of our results a much easier task.

Chapter 7 introduces two important features of the language – namely procedures and functions. These enable us to separate out parts of our program from the main body and this has the advantage that it

(1) makes the main body of the program easier to read, and
(2) means that we can test parts of the program separately from the rest,

thereby allowing us to have more confidence in the final program.

Once we have introduced these features we can fully exploit the structural approach to programming that the Pascal language encourages. This involves breaking a problem into a hierarchy of nested blocks and is sometimes called the 'top down' approach. At each stage we break units of the structure into steps which are the structural elements at the next level. Only when we can directly write the steps of a structure in the programming language do we carry out the translation. The elements of the Pascal language emphasise this structural approach: it encourages well-written programs that relate directly to the proposed solution to the problem. An immediate consequence is that it makes them easier to read and to remove any errors. It also makes them easier to write.

We therefore tackle problems by describing the steps that we propose to use in their solution, only translating them into Pascal at the last stage. Thus the techniques used for solving the mathematical problems considered in this text can be applied in any other programming language such as BASIC or FORTRAN. Pascal has the advantage over most other languages in that its programs are generally easier to follow by another reader.

Chapters 8, 9 and 10 cover an introduction to the numerical methods used to solve non-linear algebraic equations and first-order differential equations and to perform numerical integration. The proposed solutions are introduced from a heuristic viewpoint and, while there is some discussion of the errors that can occur using the various methods, the theoretical background is left to other

texts [see Phillips and Taylor (1973) or Dew and James (1983) for example].
Each technique proposed is introduced graphically and the writing of programs
to permit a solution to be obtained is covered.

One feature that becomes apparent from these chapters is how easy it is to
produce programs which allow us to tackle a wide range of numerical problems
that arise within an introductory mathematics course. The illustrations and
exercises at this stage are designed to allow the reader to explore and compare
the simple and more obvious methods available using the minimal amount of
knowledge about the programming language.

The next feature of the language that we introduce, in chapter 11, is the
array, allowing vectors and matrices to be used in our programs. An advantage
of this is that they allow us to hold large amounts of data easily in the computer's memory and open up a whole range of problems in which we can use a
computer. We begin by considering simple procedures for reading values into a
vector or matrix and carrying out simple operations on them. We then use the
new feature to tackle problems such as listing prime numbers and finding zeros
of polynomials. Chapter 12 is totally devoted to techniques for solving systems
of non-linear algebraic equations. Starting from the Gaussian-elimination method
(which is related to the way in which we are taught how to solve simultaneous
equations in school) we illustrate and suggest solutions to some of the numerical
problems that this method may generate on a computer. We also consider
alternative methods based on iterative techniques.

Pascal has many nice features, not least of which is the ability for the programmer to define his own types of variable. This is described in chapter 13 in
which we also introduce the use of simple characters and illustrate their use in
constructing displays of graphs on the terminal.

Once we can handle large amounts of data in arrays we can consider some of
the calculations and data summary methods that are covered in introductory
Statistical courses. Thus creation of frequency distributions and display of
histograms are covered as well as methods for calculating means and variances
etc. These will allow us to summarise results from simulations, an area of application and illustration in which the role of the computer is still being explored.
Thus we also cover the generation of random numbers which can be used in
programs described in the subsequent chapters.

The final part of the book introduces some of the more advanced features of
the language such as *records, pointers* and *sets*, which are not available in most
other languages, and illustrate their use in handling complex numbers, Fourier
transforms, manipulation of polynomials, queues, trees and graphs.

To summarise therefore, the text divides into two interleaved sections.
Chapters 2, 3, 6, 7, 11, 13, 16, 17 and 19 cover most of the Pascal language and
should be read in that order. Subsequent chapters build on material presented in
the earlier ones. The remaining chapters apply the programming structures so far
introduced to mathematical problems and serve both to illustrate the ways in
which the structures can be used and as an introduction to programming a

variety of techniques for solving mathematical problems. They are not therefore necessary reading material as far as learning the language is concerned, but are intended to give a sense of purpose to the learning process.

At each stage we give a number of exercises designed both to give practice in using the elements of the Pascal language and also to lead the reader into using the computer to illustrate and solve a range of mathematical problems. The reader may subsequently find it instructive to refer to Press *et al.* (1986) who provide a comprehensive survey of numerical recipes and programs for a wide range of problems, the extent of which is only touched on in this text.

Finally, it should be stressed that not all features of the Pascal language are introduced, nor is every point made about the structures covered. The complete language is covered in Jensen and Wirth (1978), who give a concise and well-written description of all the structures in the language that they created.

2 First Steps

2.1 Introduction

```
PROGRAM example1 (INPUT,OUTPUT);
VAR a,b : INTEGER; c : REAL;
BEGIN
   READ(a,b);
   c := (a+b)/2;
   WRITELN(a,b,c)
END.
```

The above is an example of a Pascal program which reads two numbers and prints their average. It is a mixture of words, numbers, punctuation and other characters.

The punctuation is critical. It forms an integral part of the syntax of the various statements and determines the way in which the computer interprets your program. It is also one of the larger sources of errors, particularly in the early days of learning to write computer programs.

The words used in the program fall into two categories:

(1) Reserved and pre-defined words. These words are part of the language and have a specific role to play. (For ease of identification we will use capitals for these.)
(2) The names (or identifiers) used to identify other quantities which have a meaning in this particular program. (We will use lower case letters for these.)

This use of capitals and lower case letters in not compulsory and most computers do not distinguish between lower case and upper case when interpreting a Pascal program. It is useful however while learning the language, as it allows us to distinguish between the words that are part of the Pascal language and the names chosen to identify quantities in a particular program.

The example program is a very simple one, but it illustrates the three necessary stages of any Pascal program. These are:

1. a heading,
2. definition of variables used in the program, and
3. the commands which actually perform the operation for which the program was written.

The layout of the text of a program is important in making it easy to read; in this example we have indented the commands that perform the calculations so that they stand out from the other parts of the program. This is typical of Pascal programs, in which we indent the parts of a program that form a particular structure so that they stand out against the statements surrounding them.

Let us begin by looking at the three parts of our example in more detail.

2.1.1 The heading

Every Pascal program must begin with a statement commencing with the reserved word PROGRAM. This is always followed by

(1) a user-chosen name which is the name of the program — our example has the name *example1* — and
(2) a sequence of words, separated by commas, contained between two brackets.

These tell the computer where to expect to find information to be read into the program and where to send information produced by the program. The minimum requirement in standard Pascal is that the word OUTPUT, which indicates that information from the program is to be sent to the terminal, is included in the PROGRAM statement. Thus, the minimal first statement in every program has the form

PROGRAM name (OUTPUT);

Any program requiring data to be entered from the terminal will also need the word INPUT, as is the case in our example program.

2.1.2 Variable declaration

The next part of the program is where we define the types of the various user-named quantities (or identifiers). All identifiers subsequently used in the program must be declared at this stage. Identifiers are strings of characters which we use as names in the program. Any number of characters can be used from the set of letters and numbers, although the first one in the name must be a letter. The only other restriction is that we should not, for obvious reasons, use a reserved word as an identifier.

Examples of valid identifier names are

mean   sum   a34   a3b4

The following though are invalid:

mean.2	because it contains an invalid character
3a	because it begins with a number
first time	because it contains a space
FOR	because it is a reserved word in the Pascal language

In our example program we have declared identifier c to be a REAL quantity while a and b are defined to be INTEGERS. We will describe these two types shortly.

2.1.3   The commands

These always start with the word BEGIN and finish with the word END. There may be more than one BEGIN . . . END block in a program but the last END must be followed by a . (full stop or period) to indicate the end of the program. The example program consists of two types of command statements. These are:

1. READ and WRITELN statements which perform the operations of reading values into the program and displaying numerical results and text from the program.
2. Assignment statements which are the ones that actually perform the calculations.

2.1.4   Statement separators – the semicolon ;

It should be noted that nearly all the lines of the program end with a semicolon. This informs the computer where one statement ends and the next begins. Pascal does not use the end of a line to indicate the end of a statement, as is common in some languages. This facility to separate statements has two advantages:

(1) more than one statement can be placed on the same line, and
(2) statements can cover more than one line, which allows us to handle long statements in a natural way.

The first enables us to place groups of short statements on a single line although we should be careful how far we carry it. For example, we could have written our first program as

```
PROGRAM  example1(INPUT,OUTPUT);VAR a,b:INTEGER;c:REAL;
BEGIN READ(a,b);c:=(a+b)/2;WRITELN(a,b,c) END.
```

However, while this alternative is compact and would work, it is not very readable, and would be difficult to understand by either another reader or yourself at a later stage. It also makes it difficult to find any errors easily.

Such a programming style is therefore not to be recommended.

Not every line ends with a semicolon and in particular the BEGIN is never followed by a semicolon. The semicolon before an END is not necessary but it is not wrong to include it. We will meet other examples, as we progress, of situations where a line must not end with a semicolon.

2.2 A Closer Look

The next step is to take a closer look at the details of our example program, describing the task of each of the statements and using them to introduce the fundamentals of the Pascal language. When we see the result of running the program, obvious improvements come to mind which allow us to introduce related concepts. As well as considering improvements in the way the program communicates with us and how the results are displayed, we will also consider ways of making the program itself more readable.

2.2.1 The user-defined variables

Our example program uses three variables *a, b* and *c*. These are declared in the second line. The first two are INTEGER variables while the third is REAL. All the variables we are going to use in the program must be declared at this stage. Failure to declare them results in a compilation error, warning us that variables are undeclared. The type of variable must be consistent with the values that are going to be defined for it at later stages in the program. Incompatible definition will result either in a compilation or runtime error.

The syntax of the variable declaration statement is

VAR list of identifiers:type;...........;

The terms in the list of identifiers are separated by commas, the type must be one of the types allowed in Pascal and the dots indicate that the previous part, list and type, may be repeated as many times as is required. The word VAR is used only once.

The possible values that a declared type can take depend on both the type and the computer being used.

INTEGER variables take the form of a whole number which can be values in a specified range such as $-2\,147\,483\,648$ to $2\,147\,483\,647$. This is an example of a *scalar type* which is a class of types in which the values a particular type takes can be described as a finite list.

REAL variables, which are not of scalar type, allow decimal parts and take values in a specified range such as -10^{78} to 10^{78}, although only the initial number of specified digits (for example, 10 or 14) are held in the computer's memory. This loss of information about numbers is a potential source of errors, as we shall see later.

The names *a, b* and *c* which we have chosen are not very informative. Since the length of an identifier name is unrestricted in Pascal, we are free to choose names that are much more informative about the actual role of each identifier in the program. We could, for example, replace *c* by any of the names, *answer*, *result* or *mean*, each of which tells us much more about the role that the identifier is playing. We could also replace *a* and *b* by *first* and *second*, say, which tells us to which of the two entered numbers they refer, although the gain here is not perhaps as great. However, making these minor changes helps to make the program more readable.

2.2.2 Data entry and display

The result of running our first program could be the following display on the terminal:

```
?3  4
        3           4 3.500000000E+00
```

The ? at the beginning of the first line would indicate a prompt by the computer that it is ready for the information, requested by the READ statement, to be supplied to the program. The form of the prompt depends on the computer and will vary between computers and compilers. The two values requested are then entered at the terminal, separated by a space, and the line is terminated and sent to the computer usually by pressing the RETURN or ENTER key.

The READ statement allows us to read values into variables. If we are entering values at the terminal, the computer will prompt us until a sufficient number have been entered and a final RETURN or ENTER issued. The standard form of the READ statement is

READ(list of identifiers);

where the identifiers, to which the values will be assigned, are separated by commas.

The output from the program is achieved by using the WRITELN statement. The form of the WRITELN statement is similar to the READ statement. It requires a list of quantities which are to be displayed. Our example shows how this can be used to display numerical values. However it can also be used to display text. Text is contained between two quotes and is included in this form among the list of identifiers. This facility to output text as well as numerical information allows us to make immediate improvements to our program.

The prompt for data, which indicates when the program is waiting for values to be entered, is not very informative since it does not tell us in any way the type and amount of data it requires at that point. We could therefore improve our program by writing it so that it gives us a reminder about the information that it is expecting at that stage. This is the first step towards making the program interact with its user, allowing the passage of information between the user and the program. In our example we can achieve this by inserting the following line immediately after the BEGIN

WRITELN(' Type in the values of two integers');

The final area in which we can make an immediate improvement is the display of the results. The output at the moment is in the form

    3   4 3.500000000E+00

which is not very informative. It also has unnecessary spaces and zeros, making it hard to read. When displaying results from a program, particularly one that is going to be used several times, we should aim to

(a) make them clear to read,
(b) make them informative, and
(c) use the available space efficiently.

Output which is either cramped or strung out is hard to read, while redundant digits and lack of descriptive information all help to make the results uninformative as well as hard to read.

We cannot really tackle the efficient use of available space here, since we need to have progressed in our programming to a stage where we can produce large amounts of numerical results. We therefore leave this problem until chapter 6. The other two objectives we can consider in terms of our example.

The display can be made more informative by adding some text to allow us to interpret the numerical values displayed. This can be simply done using text in the write statement at the appropriate places between the identifiers whose values we wish to display. Thus an improvement is achieved by replacing the original WRITELN statement by

    WRITELN('The average of ',first,' and ',second,' is ', result);

The result of this is to produce the following line of output which is clearly more informative than our first attempt.

    The average of   3 and   4 is 3.500000000E+00

Finally, we consider the way in which the numerical values are displayed.

2.2.3 Formatting output

The way a number is displayed is called its *format* and in our example we have allowed the computer to choose the format of the display of each number that is output by the WRITELN statement. In this way the display of the number depends on the type of variable in which it is stored, and the number of positions used depends on the particular compiler being used. The compiler might, for example, allocate ten positions to integer variables, filling up with blanks to the left of the number where necessary, and display real values to ten significant figures in decimal point form.

The example shows how, using the default formats, the values of integer variables are displayed with blanks filling the unused positions to the left of the number. The values of real identifiers are displayed using E format as shown by the result of printing *c*. The default is to supply the number with a single digit followed by a decimal point and nine further digits, and the letter E followed by up to two digits and a sign. In this form the number before the E has to be multiplied by 10 to the power of the number after the E. Thus if the number to the right of the E is positive, the decimal point is moved that number of places to the right, while if it is negative the decimal point is moved that number of places to the left (inserting zeros as necessary).

While the default formats are convenient for large numbers or very small real numbers, they rarely meet our requirements. Hence, an improvement is achieved in most situations if we can control the display of our results. The number of character positions used to display a number is called its *field width* and we control the display of a number by specifying its field width.

For INTEGER variables the field width simply states the number of character positions available for displaying the number. If the field width is greater than required to display the number then the field width is filled with blanks to the left of the number. If the number has more figures than allowed for in the specified field width, then sufficient extra positions will be used to output the whole number. Thus if we specified a field width of 1, then we are allowing the space allocated to that INTEGER to be determined by the magnitude of the number. This is probably adequate for most requirements except when we want output in the form of a well-laid-out table, where a little more thought is required.

The field width associated with a particular identifier is placed in the WRITELN statement immediately after the identifier's name, preceded by a colon. Thus, for example

> WRITELN(first:3);

would allocate a field width of 3 character positions for the display of the INTEGER identifier *first*.

We can also choose to display REAL numbers in E format with a different field width from the default. To achieve this a single colon and number is placed after the identifier which simply specifies the field width. The value will then be displayed with the number of figures shown equal to the field width less six. The other six spaces are taken up by the exponent information, the decimal point and any sign. Thus, for example

> WRITELN(result:10);

would display the value of the REAL identifier *result* in the form   3.500E+00.

If however we want to display the value of REAL numbers in the natural way (that is, not in the E format) as well as the field width, we must also specify the number of decimal places to be displayed. The number of decimal places used must clearly be less than the field width. To control the display of our output we supply format information in the WRITELN statement,

immediately after the variable name, in the form of two numbers each preceded by colons, the first specifying the field width and the second the number
of decimal places. For example

    WRITELN(result:4:1);

would display the value of *result* in the form   3.5 with one decimal place.

In the case of REAL values the field width must be equal to at least the
number of decimal places plus three, which allows space for the sign, at least
one term of the integer part and the decimal point. If the number does not
require the full number of positions allowed in the integer part of the number,
blanks are inserted to the left of the number. Again, if the number requires
more space than is allowed for it, then extra positions are made available for
the number to be displayed.

Examples

Suppose we have an INTEGER variable *time* having value 123 and a REAL
variable *length* having value 472.34. The following examples illustrate the
various ways of displaying that information.

The first example shows the result of using the default formats.

    WRITELN('at time',time,'length is',length);
    at time   123length is 4.723400000E2

The next example allocates five positions to the integer and a field width
of eight with three decimal places for the real value.

    WRITELN('at time',time:5,'length is',length:8:3);
    at time   123length is 472.340

The third example allocates one position to the integer which, since it is a
three digit number, takes the necessary places and the position of the rest of
the output is adjusted accordingly. The real number is allocated seven positions
and is displayed with two decimal places.

    WRITELN('at time',time:1,'length is',length:7:2);
    at time123length is 472.34

The final example illustrates the deliberate use of an E format for the real
value using twelve positions to display the number. This allows six significant
figures to be displayed, together with the sign, decimal point and the
exponent information.

    WRITELN('at time',time:1,'length is',length:12);
    at time123length is 4.72340E 02

These last two examples highlight a problem that occurs with the :1 integer
formats. The number merges with the text. To ensure separation a space must

be inserted as a character before, and sometimes after, the number is displayed. For example

WRITELN('at time ',time:1,' length is',length:12);
at time 123 length is 4.72340E 02

Similarly, to display three INTEGERS a,b,c having values 1, 2 and 3 respectively WRITELN(a:1,b:1,c:1); would result in 123 while WRITELN(a:1,' ',b:1,' ',c:1) would result in 1 2 3.

If we now apply these ideas to our example program we can improve the display of the output by replacing the line before the END statement by the line

WRITELN('The average of ',first:1,' and ',second:1,' is ',result:5:2);

which allows the size of the integer variables to determine the space and allocates five spaces to the real variable, two of which will be decimal places, a third used for the decimal point and another for a sign if that is necessary. Thus, our initial example could be written in revised form as

```
PROGRAM example1 (INPUT,OUTPUT);
VAR first,second : INTEGER; result : REAL;
BEGIN
  WRITELN('Type in the values of two integers');
  READ(first,second);
  result := (first + second)/2;
  WRITELN('The average of ',first:1,' and ',
                    second:1,' is ',result:5:2)
END.
```

The result of running this program is the following display on the terminal

Type in the values of two integers
?3 4
The average of 3 and 4 is 3.50

2.2.4 Apostrophes and new lines

We have seen that any text between quotes is written into the output display when the line is executed. However, a problem occurs when we wish to include a quote in the display. To achieve this we simply place two quotes together in the write statement. Thus to output the phrase JOHN'S GRADE we would use the statement

WRITELN('JOHN''S GRADE');

The LN after the WRITE causes the output to be displayed and a carriage return is appended, causing the next set of output to be displayed on a new line. If we simply used WRITE(. . .) then the output does not terminate with a line-end and the next set of output is added to this line. The output is only displayed once the line-end has been issued. Using WRITELN; produces a new line and displays any output stored from previous WRITE statements. If it follows a previous WRITELN(. . .); statement it produces a blank line.

The following examples illustrate the uses of these statements:

Commands	Resulting output
WRITELN('line 1');	line 1
WRITELN;	
WRITELN('line 2');	line 2
WRITE('line 1');	
WRITELN('line 2');	line 1line 2
WRITELN('line 3');	line 3
WRITE('line 1');	
WRITELN;	line 1
WRITELN('line 2');	line 2

2.3 The Assignment Statement

The only statement in our example program that we have not discussed yet is the one that actually does the mathematics. This is an example of an assignment statement. The result of the operation described on the right-hand side of the statement is stored in the variable named on the left-hand side of the statement, replacing the previous value stored in that variable.

Usually, the type of the identifier on the left-hand side of the statement must be the same as that resulting from the expression on the right-hand side. Thus, if the expression on the right-hand side of the assignment statement is a real quantity, then it must be assigned to a REAL variable. However if it is an INTEGER quantity it can be assigned to either a REAL or INTEGER variable.

We should note the use of := rather than an equals sign. This is because it is not an equality in the strict mathematical sense, but a way of indicating the value to which the identifier on the left-hand side should currently be set.

Examples

r := 3	assigns the value 3 to r
r := 0.1E−05	assigns the value 0.000001 to r
r := p	sets the value in r equal to the value currently in p
r := x + y	sets r equal to the sum of the values in x and y
r := r + x	replaces the value in r by its previous value plus the value in x

This last example illustrates the difference between a mathematical equality and the concept of assigning values to identifiers in computer programs. It shows how we can use an identifier as part of the expression that is assigned to it, even though the statement does not make sense mathematically.

2.3.1 Arithmetic operators

The expression on the right-hand side can take various forms. The simplest involves the arithmetic operations division, multiplication, addition and subtraction, denoted by the signs $/$, $*$, $+$ and $-$ respectively. In statements containing more than one operator, any division or multiplication is performed before addition or subtraction; otherwise they are carried out in the order in which they appear in the expression, reading from left to right. However, any part of the statement in brackets is performed first, and hence brackets can be used to ensure that the correct evaluation is performed. (If you are in any doubt over the result of a calculation, it is sensible to make use of brackets.)

Examples

Suppose we have four identifiers having values a:=6 b:=4 c:=2 and d:=2.

a+b/d   is equivalent to a+(b/d), giving 6+2 = 8
            that is, adding a to the result of b/d

(a+b)/d   is equivalent to (a+b)/d, giving 10/2 = 5
            that is, dividing the sum of a and b by d

a+b/d*c   is equivalent to a + (b/d)*c, giving 6 + 2*2 = 10
            that is, adding a to the result of multiplying c by the result of
            dividing b by d

a+b/(d*c)   is equivalent to a + b/(d*c), giving 6 + 4/4 = 7
            that is, adding a to the result of dividing b by d times c

Note that in mathematics ab/cd commonly means (a*b)/(c*d), whereas in Pascal and some other programming languages a*b/c*d means a*(b/c)*d.

The above operators can be used on INTEGERS or REALS. The result of division of integers is a REAL quantity and hence must be assigned to a variable that has been declared as a REAL. Failure to do so results in a compilation or runtime error.

If the expression contains both REAL and INTEGER quantities, then the result is a REAL and must be assigned to a REAL variable. Again, failure to do so will result in a compilation or runtime error.

There are also two integer operators available related to the case of division.

They are DIV and MOD which take the following form:

a DIV b   gives the integer part of (a divided by b)
                  that is, the quotient of a/b

a MOD b   gives the remainder part of a divided by b

    Note that the space between the words DIV and MOD and each of the variable names is necessary, otherwise the compiler treats the whole as an identifier which it decides is undeclared. The variables a and b must be of INTEGER type.

Examples

      i:=9 DIV 4   sets i equal to 2;   i:=24 DIV 9   sets i equal to 2
      i:=9 MOD 4   sets i equal to 1;   i:=24 MOD 9   sets i equal to 6

MOD and DIV have a higher priority than addition and subtraction: thus expressions such as a + b MOD c are interpreted as a + (b MOD c).

It should be noted that if b $>$ 0 then $-$ b MOD c is interpreted as $-$(b MOD c).
    Try to avoid using DIV and MOD with integers which can turn out to be negative. Even if n is positive, m MOD n is not necessarily between 0 and $n-1$. The rule is that the result of m DIV n is always rounded towards zero; m MOD n is the corresponding remainder. Thus

    11 DIV 4 = 2;   11 MOD 4 = 3   [2∗4 + 3 = 11]
    11 DIV (−4) = −2;   11 MOD (−4) = 3   [(−2)∗(−4) + 3 = 11]
    (−11) DIV 4 = −2;   (−11) MOD 4 = −3   [(−2)∗4 + (−3) = −11]
    (−11) DIV (−4) = 2; (−11) MOD (−4) = −3 [2∗(−4) + (−3) = −11]

2.4 Standard Functions

The operation on the right-hand side of the assignment statement can also make use of various standard functions available in the language.

Examples

    x := COS(a)
    y := 3 ∗ SIN(a) + COS(b) / 4
    x := 4 ∗ LN(y)   sets x equal to 4 times the natural logarithm of y
    y := EXP(z)   sets y equal to e^z

Note that the trigonometric functions work in radians.

There is no direct way of raising a number to a power in Pascal, but we can make use of the standard functions EXP and LN to achieve this. Thus x to the power 3.4 can be achieved by the operation EXP(3.4*LN(x)). Clearly though, if x is negative the LN(x) will result in an error. Integer powers can be more efficiently calculated using SQR. For example

x^4 by SQR(SQR(x)) x^3 by x*SQR(x) and $(-2)^2$ by SQR(-2)

2.4.1 Arithmetic functions

A list of functions available in Pascal is given in appendix A. The functions available to perform various arithmetic operations can be classified into three types, as shown in the following examples. Note carefully the remarks about the types of the arguments and the results.

Functions producing results of the same type as the argument

SQR(x)   which produces the square of the number x. If x is real, the result is
real while if x is an integer, the result is an integer

ABS(x)   gives the value of $|x|$, that is the absolute value of x. For example,
y := ABS(-3) and y := ABS(3) both assign the value 3 to y.
Again ABS(x) has the same type as x.

Functions producing real results irrespective of the argument type

SQRT(x)   which produces the square root of x
(clearly x must have a positive value, otherwise an error results)

SIN(x)   which produces the sine of the angle x radians

LN(x)   which gives the logarithm of x to the base e.
(x must have a positive value, otherwise an error results)

Functions producing integer results from real arguments

ROUND(x)   produces the integer result of rounding the real value x to the
nearest integer. For example

y := ROUND(21.4) sets y equal to 21
y := ROUND(21.5) sets y equal to 22
y := ROUND(20.6) sets y equal to 21
y := ROUND(-21.6) sets y equal to -22
y := ROUND(-21.5) sets y equal to -22

TRUNC(x)   gives the integer resulting from truncating x. That is simply
removing the decimal part. For example

y := TRUNC(31.7) sets y equal to 31
y := TRUNC(−31.7) sets y equal to −31

2.5 Constants

There are numerous instances when we need a constant which has some definite meaning or is used many times throughout a program. By declaring the value as a constant before we enter the main block of the program, we can make the program easier to read and to modify without using a variable name. The identifiers which are constants are defined and assigned their numerical values before any variables are defined. The declaration is done by the command

CONST identifier=value; . . . .;

where again the dots indicate that more than one constant can be defined. Note the use of 'equals' and not :=. The word CONST, like the word VAR, can only occur once in each segment of the program.

The following example illustrates the use of a declared constant for storing an approximation for π.

```
END.
PROGRAM circle (INPUT,OUTPUT);
(* This program illustrates the use of a declared
constant and a comment *)
CONST pi = 3.14159;
VAR area,radius,circumference : REAL;
BEGIN
   radius := 4;
   circumference := 2 * pi * radius;
   area := pi * radius * radius;
   WRITE('a circle of radius ', radius:3:1);
   WRITE(' has circumference ', circumference:6:2);
   WRITELN(' and area ', area:6:2)
END.
```

The output from this program would be the following single line:

a circle of radius 4.0 has circumference 25.13 and area 50.27

It should be noted that to ensure a space between a numerical value and any subsequent text, a space must be placed at the beginning of the text.

2.6 Comments

In the above example we also inserted an extra line, which formed no part of the Pascal code, but does remind us of the objective of the program. Comments of this kind are very useful both for this purpose and for recording the exact purpose of the various identifiers in the program. Comments may be inserted at any stage in the program by including them between either (∗ and ∗) or {and}.

It is particularly important to include the closing parenthesis of the comment, otherwise part of the program may be intepreted as part of the comment. This is because all the text up to the next close of comment would be thought to be part of the first comment. This can have strange side-effects since part of our program would apparently be missing. Any error messages that resulted would not usually indicate the problem directly so if unexplained error messages occur, it is always worth a quick check that all comments are terminated at the correct place.

2.7 Compiling and Running a Program

To complete this introductory chapter we need to consider the problem of running the program on a computer. The specific details obviously depend on the computer being used and the Pascal compiler that is available on that machine.

The first step is to enter the text of the program into the computer's memory and to save it as a file on a device available for storing programs.

The next step is to compile the program. This is a process which uses another program, known as the compiler, to check the syntax of the program and convert it, from the Pascal language that we can read, into machine code that can be run on the computer. There is no need for the user to see or understand the translated version of the program, although it can often be saved in a file, so that we can avoid having to recompile a working program every time we want to run it.

The compiler scans through the program and should, among other things, check that

(1) every variable used is declared,
(2) the various constructions are correct,
(3) semicolons are present where expected,
(4) each BEGIN has an associated END,
(5) assignment statements are valid, for example
    (a) all brackets that are opened have a close,
    (b) the resulting type of an expression is assigned to a correct type of identifier,
    (c) standard functions have the correct number and type of parameters,
(6) comments and text in WRITE or WRITELN statements have a beginning and an end.

If an error is detected, the line number and the position on the line where the detection occurs is usually displayed, together with an error number. Some compilers also give a brief description of the error.

The actual error can be found at or before the point indicated. It should be noted that this can mean that the error could have occurred in the lines before the line indicated.

Another slightly disconcerting problem is that subsequent errors may be caused by an earlier error. Hence a useful rule to follow when removing errors from a program is to start with errors at the top of the program and work down, recompiling after the obvious ones have been removed at each stage.

We illustrate these points by deliberately making two errors in our first program. Suppose we had omitted a semicolon at the end of the variable definition statement and instead of typing

result := (first + second)/2;

we had typed

result := (fist + second)/2;

Thus our program reads as follows with the errors included.

```
PROGRAM example1 (INPUT,OUTPUT);
VAR first,second : INTEGER; result:REAL
BEGIN
  WRITELN('Type in the values of two integers');
  READ(first,second);
  result := (fist + second)/2;
  WRITELN('The average of ',first:1,' and ',
                    second:1,' is ',result:4:2)
```

If we now compile this code, then something similar to the following may be displayed.

```
PROGRAM example1 (INPUT,OUTPUT);
VAR first,second : INTEGER; result : REAL
BEGIN
  WRITELN('Type in the values of two integers ');
$
** ERROR 5 at line 4.      ';' expected.
  READ(first,second);
  result:= (fist + second)/2;
                  $
** ERROR 1 at line 6.    Variable identifier expected.
  WRITELN('The average of ',first:1,' and ',
                      second:1,' is ',result:5:2);
                                                $
**   ERROR  159  at  line  8.    Illegal operation on
  these operands.
  END.
```

The first error message indicates that it was detected at the beginning of line 4. Clearly a semicolon cannot be placed at the beginning of a line, so the error must have occurred earlier. A semicolon cannot be placed after a BEGIN, which leads us to the end of line 2 where the error occurred.

The second error was detected at the + sign in line 6 and the message indicates that it was expecting a declared identifier at this point. This is caused by the fact that *fist* had not been declared. The identifier *first* was clearly intended here.

The third error in line 7 indicated that an illegal operation was being carried out. It is, in fact, the attempt to display the value of *result* which because of the second error will not have been assigned a value. Hence it cannot be used in any subsequent statement. This error is a direct result of the second error and therefore requires no further action.

The problem of subsequent errors varies tremendously from one compiler to another, and problems such as the third error above will not always be reported.

Once we have successfully compiled the program (that is, the compilation produces no error messages) we can proceed to run the program.

The result of running a program will be either:

1. A successful run, producing the required result from the program, or
2. A successful run, producing an incorrect result, or
3. An unsuccessful run, producing a list of runtime errors, or
4. A successful run, producing what is thought to be the required result from the program.

Again, error messages are displayed relating to particular lines but the actual error may have occurred much earlier in the program.

For example, suppose we had a division operation in an assignment statement and at the point the statement is executed the divisor contains the value zero. Then an error message such as

ERROR   Division by zero at line 6 in example1.

would have resulted, where *example1* is the name of the part of the program in which the error occurred.

While many runtime errors are easy to sort out, many require an element of detective work to unravel. The necessary skills can only be developed with practice and experience.

2.8 Exercises

2.1. Which of the following are valid Pascal identifier names giving reasons for those that are not allowed.

positive, word, 3times, begin, time-of-arrival,
number1, program, no of observations

2.2. Enter and run the following Pascal program on your computer.

```
PROGRAM logtest (INPUT,OUTPUT);
VAR x : REAL;
BEGIN
  WRITELN('Type in a real number');
  READ(x);
  WRITELN('The value of log ',x,' is ',LN(x));
END.
```

Run the program four times entering values of 0.5, 1, 52 and −1 when requested to type in a real number and explain what you observe.

2.3. Edit the program in exercise **2.2**, replacing 'the value of log' by 'the value of exponential' and LN(x) by EXP(x).

Run the new program four times with values of −10, 4.37, 200 and −250 respectively and explain what you observe.

2.4. If a, b and c are of type REAL and i, j, k are of type INTEGER, which of the following statements are not allowed and why not.

$$a := i/j; \qquad c := a + i/b; \qquad k := i/j;$$
$$c := i\,MOD\,j; \quad b := a\,MOD\,k; \quad k := i\,DIV\,j;$$

2.5. Examine the following program.

```
PROGRAM ARITH (INPUT,OUTPUT);
VAR a,b,c : REAL;
    i,j,k : INTEGER;
BEGIN
  READ(b,c,j,k);
  a := b + c;
  WRITELN('a = ',a);
  a := a + b / c;
  WRITELN('a = ',a:6:2);
  i := j DIV k;
  WRITELN(' i = ',i);
  a := j / k;
  i := j MOD k;
  WRITELN(' a = ',a:6,' i = ',i:4);
  WRITELN(' i = ',i:1);
END.
```

If the values b=24, c=7, j=43 and k=8 are entered, what should the output from the program look like.

2.6. What are the results of the following assignments.

$$a := 3 + 4/8; \quad a := (3 + 4)/8; \quad a := 3/4 + 8; \quad a := 3/(4 + 8)$$

2.7. Write a program that reads two integers and reports their sum and product in the form of two equations. For example, the output from the program should be of the form

$$7 + 6 = 13$$
$$7 * 6 = 42$$

2.8. Write a program that will read in a temperature in Centigrade and display the corresponding fahrenheit temperature using the conversion formula

$$F = 32 + (9/5) C$$

2.9. Write a program to read in the lengths of the three sides of a triangle, *a, b* and *c*, and display the area of the triangle using the formula

$$\text{area} = \sqrt{\{s\,(s-a)(s-b)(s-c)\}}$$

where *s* is half the perimeter, and the value of $\cos(A)$ using the cosine formula $\cos(A) = (b^2 + c^2 - a^2)/2bc$.

2.10. Write a program to read in the value of a REAL variable x and then display the values of

$$\sinh(x) = (e^x - e^x)/2,$$
$$\cosh(x) = (e^x + e^x)/2 \text{ and}$$
$$\tanh(x) = \sinh(x)/\cosh(x)$$

3 Programming Structures

3.1 Designing a Program — the Algorithm

An algorithm can be defined as a procedure for solving a problem in a finite number of steps. There are usually many procedures which can be followed in order to produce a solution to a problem. The objective in programming a computer is to produce the most efficient and accurate solution. This is not necessarily the easiest or the most natural solution to the problem.

Suppose, for example, we wanted to use a calculator to evaluate the quadratic function $ax^2 + bx + c$, then we could follow the steps:

1. evaluate the square of x
2. multiply it by the value of a
3. store the result
4. multiply b by x
5. add this to the stored result
6. add c to the stored result.

This is not unique since we could also proceed as follows:

1. multiply a by x
2. add b to the result
3. multiply the result by x
4. add c to the result

which is equivalent to evaluating $(ax+b)x + c$.

The first is probably the natural way for a mathematician to view the problem, whereas a computer scientist should use the second. They both produce the same result, but the latter is more efficient since it achieves its aim in two multiplications and two additions. The first method uses three multiplications and two additions as well as having to store a partial result on the way.

A computer would perform the second operation more efficiently since it requires fewer operations. For the problems we are going to be dealing with at these early stages, such fine points are not critical but perhaps worth bearing in mind.

The main point is that the problem should be broken down into a set of easy-to-follow steps which we can then translate into Pascal code for use on the computer. By structuring our problems in this way we will find it easier to write

25

a program that solves our problem. We should also be able to produce a program that can be understood at a later date since the structure of the original problem should be contained in the program used to solve it.

Suppose we consider the following simple problem. An object is thrown vertically upwards with velocity u. We know that after a time t its height is given by the equation

$$h = ut - gt^2/2$$

where g is a constant representing the acceleration due to gravity. We can determine the times when it is at a given height (once on the way up and once on the way down) by solving the quadratic equation, the solution being

$$t = \frac{u \pm \sqrt{(u^2 - 2gh)}}{g}$$

How can we write a program to perform this operation for us?

We begin by breaking the task down into the following steps:

1. read in the initial velocity (u) and the height (h)
2. calculate the root discriminant, $\sqrt{(u^2 - 2gh)}$
3. calculate the first time: $(u - \text{root discriminant})/g$
4. calculate the second time: $(u + \text{root discriminant})/g$
5. write out the results.

By laying the steps out in this way we have a basic structure for the eventual program and an indication of the identifiers that it will require. The names that suggest themselves are

gravity, which we will define as a constant
and velocity, height, time1, time2, rootdiscriminant and discriminant.

These latter terms are variables which take on real values since they are used in calculations which involve the use of square roots and division. It is also the natural way in which we think of the quantities that they represent. Hence they should be of type REAL.

This choice of names for identifiers, relating to their use, has the advantage that if we return to the program at a later date, we immediately have an indication of the role that the various identifiers play.

Using the small number of features of the Pascal language that we have so far covered, we can translate the information in the above into Pascal code.

```
PROGRAM example2 (INPUT,OUTPUT);

(* This is a program to calculate the time taken to
reach a given height by a particle projected vertically
upwards.
    Users are requested to supply the initial velocity
and the height in which they are interested.  *)
```

```
CONST gravity = 9.81;
VAR velocity,height,time1,time2 : REAL;
    rootdiscriminant,discriminant : REAL;

BEGIN
(* Step 1   data entry section *)
   WRITELN('What is your initial velocity',
                             ' in metres per second?');
   READ(velocity);
   WRITELN('What is the height in metres',
                     ' in which you are interested ?');
   READ(height);

(* Step 2   calculate the root discriminant*)
   discriminant := velocity * velocity -
                               2 * height * gravity;
   rootdiscriminant := SQRT(discriminant);

(* Steps 3 and 4   calculate the two times *)
   time1 := (velocity - rootdiscriminant)/gravity;
   time2 := (velocity + rootdiscriminant)/gravity;

(* Step 5 display the results *)
   WRITELN('at an initial velocity of ',
                     velocity:6:2,' metres per second');
   WRITELN('the particle is at height ',
                     height:6:2,' metres');
   WRITELN('at times ',time1:6:2,' and ',
                     time2:6:2,' seconds.');

END.
```

The result of running this program is the following display on the terminal:

> What is your initial velocity in metres per second ?
> ?25
> What is the height in metres in which you are interested ?
> ?30
> at an initial velocity of 25.00 metres per second
> the particle is at height 30.00 metres
> at times 1.93 and 3.16 seconds.

This is not however the end of the problem, since if we ran the program as above but requested a height of 100 then the term supplied as an argument to the function SQRT is negative and produces the runtime error message as follows:

> What is your initial velocity in metres per second ?
> ?25
> What is the height in metres in which you are interested ?
> ?100
>
> ERROR   SQRT of a negative number attempted line 17 of example2.

The actual form of the error message depends on the compiler and the implementation of Pascal that is being used. In some cases the error messages are confusing and can be difficult to interpret, the exact meaning or cause of the problem only being established with careful and, in some cases, prolonged study of the program.

It would be useful to anticipate this problem and build into our program a means by which to avoid it. Thus, we need to ensure that the calculations are performed only if the conditions for a real root to exist are satisfied. This involves building a conditional statement into the program which tests whether the argument is non-negative and avoids the calculations when it is not. This is an example of a situation where we need a statement which controls whether or not other statements are obeyed, and if so how many times.

3.2 Control Statements

To tackle the problem posed by our second example program we need to introduce statements that allow us to control the operations to be carried out. The example programs we have met so far have all consisted of sets of statements obeyed in the order in which they are listed. In our example we require the ability to test whether the quantity whose square root we are going to calculate is positive and only proceed to perform the calculation if that is the case.

Another area in which we could usefully exercise some control over the order in which the statements are obeyed is that in which repetitive operations are required, such as in calculating sums. This could be achieved by simply repeating the statements; but while this may be satisfactory for programs attempting small tasks, it can soon become tedious and very restrictive.

There is therefore a real need for statements which control the operations to be carried out; so we begin by considering the IF statement, which allows us to add the protection to our example program so that it will not produce an error.

3.2.1 The IF statement

The IF statement allows us to perform the next block of statements only if a condition is satisfied. It can also be used to allow us to choose between two blocks of statements, depending on whether a condition is true or false.

The syntax of the two forms of the statement are:

(1) IF condition THEN
         BEGIN
            set of statements
         END;

In this case the set of statements is obeyed only if the condition is true.

(2) IF condition THEN
      BEGIN
          first set of statements
      END
    ELSE
      BEGIN
          second set of statements
      END;

In this case the first set of statements is obeyed if the condition is true, otherwise the second set of statements is obeyed since the condition is false. Single statements need not be surrounded by a BEGIN and END, but it should be noted that since the IF-THEN-ELSE is considered to be a single statement the ELSE must not be immediately preceded by a semicolon which is treated as a statement separator.

The condition in the IF statement is a logical or BOOLEAN statement which can be either true or false. It can take various forms, many of which make use of the relational signs

>   greater than
<   less than
=   equals
$\geqslant$ greater than or equal to
$\leqslant$ less than or equal to
$\langle\rangle$ not equal to

which can be used to form a condition relating two identifiers or expressions. These can be REAL or INTEGER quantities but, because of the inaccuracies inherent in all arithmetic carried out on REALS, resulting from the way the computer rounds and stores REAL values, = and $\langle\rangle$ should only be used with INTEGER expressions.

Examples

alpha $<$ 3.4	true if alpha is less than 3.4
i = j	true if i has the same value as j
alpha $>$ beta	true if alpha is greater than beta
i $>=$ (j DIV k)	true if i is greater than or equal to the integer part resulting from dividing j by k

If x is a REAL variable, we cannot sensibly check that its value equals a number since its value is rounded when stored in the computer. We can only check that it is sufficiently close to that number — that is, it is accurate to so many significant figures. For example, the condition

$$\text{ABS}(x-4) < 1\text{E}-5 \quad \text{or alternatively} \quad \text{ABS}(x-4) < 0.00001$$

tests whether x is within 0.00001 of the value 4 (that is, it is accurate to 4 decimal places).

The logical operators NOT, AND and OR can be used to build other conditions. These are used as follows

> NOT condition
> condition1 AND condition2
> condition1 OR condition2

and operate on the conditions in the following way:

> NOT (condition) negates the conditions and is true if the condition is
> false and false if it is true

The other two operations are best described in truth tables.

The AND statement is true only if both conditions are true as follows:

condition1	*condition2*	*condition1 AND condition2*
true	true	true
true	false	false
false	true	false
false	false	false

The OR statement is true if either of the two conditions is true or both of them are true as follows:

condition1	*condition2*	*condition1 OR condition 2*
true	true	true
true	false	true
false	true	true
false	false	false

If relational statements are used in these conditions then they must be contained in brackets.

Examples

> (alpha $<$ beta) AND (i $>$ = 6)

is true only if both conditions are satisfied.

> (i $<$ k) OR (i $<$ = j)

is true if either or both conditions are satisfied.

> NOT((i $<$ j) OR (i $>$ k))

is true only if neither condition is satisfied.

In a compound statement order of priority is NOT, AND and finally OR with priority given to those parts of the statement contained in brackets.

3.2.2 BOOLEAN variables

It is useful at this stage to introduce a new type of variable, the BOOLEAN, which can only take values TRUE or FALSE. BOOLEAN is another example of a *scalar* type since it can only have one of a set of values which can be listed. In this case it is a finite set consisting of the two values TRUE and FALSE.

An identifier of type BOOLEAN, provided that its value has been set earlier in the program, can be used as the condition in the various statements that require one.

Example

VAR convergence : BOOLEAN;   Declares that *convergence* is a BOOLEAN
·   variable. This statement occurs in the
·   declaration block at the top of the program
·

convergence := TRUE;   Sets *convergence* to have the value TRUE
·
·

IF convergence THEN . . . . . .   Uses *convergence* as a condition which, since
                             it is true, means that the set of statments
                             following the THEN are obeyed

The value of a BOOLEAN variable can also be set by assigning a relational condition to it. For example

positive := i > 0;   assigns the value true to *positive* if i is greater than 0
                     otherwise sets it to false

This is a far more efficient way of setting the value of a BOOLEAN variable than using the following construction based on the IF statement:

IF i > 0 THEN positive := true
        ELSE positive := false;

Finally, there are standard functions which produce a BOOLEAN result; one of these is the function ODD(x) which has the value TRUE if the integer variable x has an odd value. We shall meet other BOOLEAN functions later on.

3.2.3 Protection of programs

One use of the IF statement is to protect parts of the program from invalid data or values that can cause a problem. We can use an IF statement to protect functions from invalid arguments, such as the square root function in our

example program for calculating the time to reach a given height of the projectile. The following is an illustration of a much better program since it
anticipates a possible problem. The additional statements are in italics.

```
PROGRAM example2mark2(INPUT,OUTPUT);
CONST gravity = 9.81;
VAR velocity,height,time1,time2:REAL;
    rootdiscriminant,discriminant:REAL;

BEGIN

  WRITELN('What is your initial velocity',
                          ' in metres per second ?');
  READ(velocity);
  WRITELN('What is the height in metres',
                  ' in which you are interested ?');
  READ(height);

  discriminant:=velocity*velocity-2*height*gravity;
  IF discriminant >= 0 THEN
      BEGIN
          rootdiscriminant := SQRT(discriminant);
          time1 := (velocity - rootdiscriminant)/gravity;
          time2 := (velocity + rootdiscriminant)/gravity;
          WRITELN('at  an  initial  velocity  of ',
                    velocity:6:2,'  metres  per  second');
          WRITELN('the particle is at height ',
                            height:6:2,'metres');
          WRITELN('at times ',time1:6:2,' and ',
                            time2,:6:2,' seconds');
      END
  ELSE
          WRITELN('This height is unattainable',
                      'with your initial velocity');
END.
```

3.2.4 *The CASE statement*

The IF statement allows us to select either of two alternatives, and although it
can be used to choose between more the constructions can become lengthy and
complex. The Pascal language, however, has another construction which allows us
to select one of a set of alternatives. It is the CASE statement.

The general form of the CASE statement is:

```
CASE scalar-valued expression OF
        set of commands of the form
              label1,label2, .... labeln:statement
    END;
```

The labels are values that the expression can take and the statement is any valid PASCAL statement. (Note that, through using a BEGIN and END block, the statement can be a set of statements.)

When the CASE statement is reached, the value of the expression is calculated and then the statement proceeded by that label is obeyed. The value of the expression must take one of the labels currently in the list or a runtime error occurs. The scalar-valued expression therefore acts as a selector which decides which of the choices is used.

Examples

The first example shows how to select any of seven output statements, depending on the value of the INTEGER variable day.

```
CASE day OF
    1 : WRITELN('Sunday');
    2 : WRITELN('Monday');
    3 : WRITELN('Tuesday');
    4 : WRITELN('Wednesday');
    5 : WRITELN('Thursday');
    6 : WRITELN('Friday');
    7 : WRITELN('Saturday');
END;
```

This would type the day of the week according to the current number assigned to the identifier day. Thus, for example, if day had the value 6 then the word 'Friday' would be displayed.

The second example shows how the same statement can be obeyed for different values of the INTEGER variable *time*. It also illustrates the use of a blank statement when we want to do nothing for certain values of the selector.

```
CASE time OF
    1,5 : x:=x+5;
    2,3 : x:=x+3;
    4   : ;
    6   : x:=x-20;
END;
```

This statement would add 5 to the value of x if the INTEGER *time* has the value 1 or 5, add 3 if *time* equals 2 or 3, do nothing if *time* has the value 4 and subtract 20 if *time* has the value 6. Note that any other value of *time* would cause an error.

The above examples illustrate the use of the case statement to enter and display values of user-defined types. It can also be used to perform sets of operations conditional on values taken by the expression.

Thus, for example, if *convergence* is a BOOLEAN variable which is TRUE only if an iterative process has converged (we will discuss this problem in chapter 8), then to display the results we could use the following construction:

```
CASE convergence OF
   TRUE :
      BEGIN
         WRITELN('convergence successful after ',
                                      iterations:5);
         WRITELN('the solution is ',x:5:2);
      END;
   FALSE :
      BEGIN
         WRITELN('convergence unsuccessful');
         WRITELN('final value is ');
      END;
END;
```

While this is achieved equally well using the IF construction considered first, the CASE statement would of course allow us to use more than two alternatives without constructing nested IF statements.

3.3 Looping

We have already suggested the need for constructions that allow us to repeat statements or to loop round a part of the program. The requirements in mathematics to calculate sums of quantities or to repeat a set of operations until a condition is satisfied strengthen this need. There are two types of looping that we need to consider. The first is the ability to repeat a set of operations a pre-determined number of times, such as calculating the mean of a set of numbers. Here we can usually calculate beforehand how many terms we must add together, and we simply need the facility to repeat the operation. The second occurs when we do not know how many times we need to repeat the operation before we enter the loop. For example, we may want to accumulate a sum until it exceeds a particular value — in which case we clearly do not know at the outset the number of times we must repeat the operation.

3.3.1 The FOR loop

Let us consider the simple problem of summing four numbers entered at the terminal. Using the facilities that we have available so far, we could use either of the following two programs

```
PROGRAM example3 (INPUT,OUTPUT);
VAR sum,next : REAL;
BEGIN
    sum := 0;
    READ(next);
    sum := sum + next;
    READ(next);
    sum := sum + next;
    READ(next);
    sum := sum + next;
    READ(next);
    sum := sum + next;
    WRITELN(sum:6:2)
END.

 PROGRAM example4 (INPUT,OUTPUT);
 VAR a,b,c,d,sum : REAL;
 BEGIN
    READ(a,b,c,d);
    sum := a + b + c + d;
    WRITELN(sum:6:2);
END.
```

Both of these complete the task; the second in a neater way than the first, but they are both inflexible. If we wanted to use the program to sum a larger number of terms they would both need to be rewritten to accommodate the extra terms.

The first example achieves its aim by repeating the same operation a deterministic number of times. Pascal has various command structures which allow us to do this in a neater way. The simplest of these is the FOR loop. This is a construction which allows us to repeat a set of statements a pre-determined fixed number of times. Its syntax is either

> FOR identifier := expression1   TO   expression2   DO
>
>> BEGIN
>>> Sequence of statements obeyed as the value of the identifier
>>> increases from the value resulting from the first expression
>>> to that of the second in steps of size 1.
>> END;

or

> FOR identifier := expression1   DOWNTO   expression2   DO
>
>> BEGIN
>>> Sequence of statements obeyed as the value of the identifier
>>> decreases from the value resulting from the first expression
>>> to that of the second in steps of size 1.
>> END;

The following points should be noted:

(1) The identifier must be of scalar type, such as INTEGER, and is called the control variable. It increases or decreases during the execution of the loop, but its value is undefined once we exit from the loop. Furthermore, its value cannot be altered within the loop by using an assignment statement.
(2) The expressions used must be of same scalar type as the identifier.
(3) If only a single statement is to be repeated, the BEGIN and END indicating the block are not necessary.

Using FOR loops we could now rewrite the first of the above programs as follows:

```
PROGRAM example3mark2 (INPUT,OUTPUT);
VAR sum,next : REAL; i : integer;
BEGIN
   sum := 0;
   FOR i := 1 TO 4 DO
      BEGIN
         READ(next);
         sum := sum + next;
      END;
   WRITELN(sum:6:2);
END.
```

It is now a very simple task to modify this to allow us to use the program to sum any number of terms.

```
PROGRAM example3mark3 (INPUT,OUTPUT);
VAR sum,next : REAL; i,n: integer;
BEGIN
   sum := 0;
   WRITELN('How many terms do you want to sum ?');
   READ(n);
   FOR i:= 1 TO n DO
      BEGIN
         READ(next);
         sum := sum + next;
      END;
   WRITELN(sum:6:2);
END.
```

The control variable takes values which can be used but not altered in the loop. Thus, to sum the values of the first 10 integers we can use the following:

```
n := 10;
sum := 0;
FOR i := 1 TO n DO
   sum := sum + i;
```

Note how the variables *sum* and *n* have values assigned them before the loop is entered. If this was not done then a runtime or compilation error of undefined value should result. There are, however, compilers which allow identifiers to assume whatever value is currently residing in the appropriate part of the computer's memory, which can lead to very strange results.

The DOWNTO in the FOR statement is used when the value of expression1 is greater than the value of expression2 and you wish the counting variable to decrease.

If the number of steps possible in a FOR loop is negative — that is, expression1 is greater than expression2 in an increasing loop or less than expression2 in a decreasing loop — then the loop is skipped and its statements are ignored. For example

```
FOR i := 5 TO 2 DO
      WRITELN(i);
```

would result in the WRITELN statement never being obeyed.

3.3.2 REPEAT and WHILE loops

The types of control statements that we have considered so far are programming structures which perform a set of statements a pre-determined number of times. Once in the case of IF statements and many times (the number of steps) in the case of the FOR loop. It is not always the case that we will know beforehand how many times a loop needs to be repeated. There will also be many occasions when we would like to repeat a section of the program until a condition is satisfied.

There are two forms of loop which allow us to do this. They are the REPEAT loop and the WHILE loop, both of which allow us to perform the statements in the loop until a condition is satisfied — that condition being tested on each pass through the loop. Unlike the other structures that we have considered, a danger with these two loops is the ease with which a loop can be created which never terminates. To avoid this, we must ensure that the truth of the condition will eventually alter in such a way as to terminate the loop. The conditions used in these loops take the same forms as the conditions used in the IF statement.

The REPEAT loop takes the form

```
REPEAT
      set of statements
UNTIL condition;
```

The set of statements is obeyed at least once since the condition is tested at the end of the loop. The set of statements is repeated until the condition becomes true. Since the REPEAT loop has a terminating statement, there is no need to bracket the set of statements between a BEGIN and an END.

The WHILE loop takes the form

WHILE condition DO
    BEGIN
        set of statements
    END;

In this case the set of statements is obeyed while the condition is true. Once the condition is false when it is tested, the operation of the program moves on to the next statements. Since the condition is tested at the beginning of the loop, the loop is not executed at all if the condition is false. If the WHILE loop consists of a single statement, the BEGIN and the END can be omitted.

The difference between the two loops is therefore that the REPEAT loop is executed at least once since the condition is tested at the end of the loop, while the WHILE loop is not necessarily executed since the condition is tested at the start. Since the test for a WHILE loop takes place at the beginning, the variables that it contains must have values assigned to them before the loop is entered. It is not necessary for the condition in the REPEAT loop to be assigned a value before entering the loop, as the value can be assigned and altered during the loop.

Examples

We illustrate the use of the two loop constructions in the following two programs both of which could be used to calculate

$$f(n) = 1 + 1/2 + 1/3 + 1/4 + \ldots + 1/n$$

```
PROGRAM example5(INPUT,OUTPUT);
VAR n : INTEGER; f : REAL;
BEGIN
  WRITELN(' enter the value for n');
  READ(n);
  WRITE(n);
  f := 0;
  WHILE n > 0 DO
    BEGIN
      f := f + 1/n;
      n := n - 1;
    END;
  WRITELN(f);
END.


PROGRAM example6(INPUT,OUTPUT);
VAR n : INTEGER; f : REAL;
BEGIN
  WRITELN('enter the value for n');
  READ(n);
  WRITE(n);
  f := 0;
```

```
      REPEAT
        f := f + 1/n;
        n := n - 1;
      UNTIL n = 0;
      WRITELN(f);
END.
```

3.4 Nesting of Control Statements

Any control statement can be included as one of the statements that is obeyed within any of the control statements, provided that it is completely contained. This is because all of the lines that make up a particular control statement are considered as a single statement. We then say that one control loop is nested within the other. We can see the set-up by considering the following two examples based on FOR statements.

```
FOR i := 1 TO 3 DO
      FOR j := 1 TO 2 DO
                   WRITE(i,j);
```

would produce the output

  1 1 1 2 2 1 2 2 3 1 3 2

from which we note that the values of the inner control variable are changing faster than the outer control variable. This is an example containing a single statement and, since there is no BEGIN or END indicating a block for either FOR loop, the semicolon at the end terminates both statements.

  The second example illustrates a more complex situation although it behaves in the same way.

```
FOR i := 1 TO numberofsamples DO
   BEGIN
      READ(samplesize);
      FOR j := 1 TO samplesize DO
          BEGIN
             READ(next);
             mean := mean + next;
          END;
      mean := mean/samplesize;
      WRITELN('Mean of sample ',i:1,' is equal to ',
                                    mean:6:2);
   END;
```

  REPEAT and WHILE loops can also be nested following the same simple guidelines.

  IF statements can also be nested within loops again, provided that they are completely enclosed in the loops. A little care is required when nesting IF statements, particularly in cases involving the use of an ELSE part to the statement. The first example is fairly straightforward.

```
IF  a  <  b  THEN
        IF  b  <  c  THEN
                WRITELN('a  <  b  <  c')
        ELSE
                WRITELN('a  <  b  but  b  >=  c')
ELSE
        WRITELN('a  >=  b');
```

In this example there is an ELSE term associated with each IF. Notice that there are no semicolons immediately before the ELSE phrases. The second example contains only a single ELSE statement which is associated with the first IF statement.

```
IF  a  <  b  THEN
        BEGIN
            IF  b  <  c  THEN
                    WRITELN('a  <  b  <  c');
        END
ELSE
        WRITELN('a  >=  b');
```

Notice that we place the inner IF statement within a block to ensure that the ELSE statement is associated with the correct IF. It is often necessary with IF statements to block parts of the program to make it clear, as in this case, which statements go together. The BEGIN and the END indicate that the second IF statement is a complete block, implying that the ELSE relates to an earlier IF.

Failure to include the BEGIN and END in this particular example results in the following piece of code:

```
IF  a  <  b  THEN
        IF  b  <  c  THEN
                WRITELN('a  <  b  <  c')
        ELSE
        WRITELN('a  >=  b');
```

This would do nothing if $a \geq b$ and incorrectly state that $a \geq b$ if $a < b$ but $b \geq c$.

Careless formulation of sequences of IF statements can be very costly in computing terms, and it is worthwhile attempting to order the conditions in some decreasing order of their likelihood of being true. Thus, for example, if we have n mutually exclusive conditions $c_1 \ldots c_n$ each instigating a distinct action s_i $i=1 \ldots n$, then they should be ordered such that if $P(c_i)$ is the probability of c_i being true then

$$P(c_1) > P(c_2) > \ldots > P(c_n)$$

Once they have been ordered in this way, then they can be tested using the following sequence of IF statements.

```
IF c1 THEN s1
    ELSE   IF c2 THEN s2
        ELSE .....
                .................
                ELSE IF c(n-1) THEN s(n-1) ELSE sn;
```

Once the condition that is true is reached, then the appropriate set of statements is obeyed and the remainder of the IF statements is bypassed.

3.5 Exercises

3.1. For a projectile projected at an angle θ to the horizontal with an initial velocity v_0, the range is given by

$$R = v_0^2 \sin(2\theta)/g \quad \text{where } g = \text{acceleration due to gravity}$$

the maximum height reached is given by

$$H_{max} = v_0^2 \sin^2\theta/2g$$

and the time of flight is

$$T = 2v_0 \sin\theta/g$$

Write a program to request the initial value and angle of projection and output the values for the range, maximum height and time of flight.

3.2. The period of a small angle swing for a pendulum of length l is given by

$$P = 2\pi \sqrt{(l/g)} \quad \text{where } g \text{ is the acceleration due to gravity}$$

Write a program which calculates either
(a) the period for a given length or
(b) the length for a given period.

3.3. Write a program that reads in an integer value n and prints out the first 12 terms of the n times table, provided n is less than 12.

Modify the program to produce the first 12 terms of all the times tables from 1 up to n.

3.4. Write a program to display the values of $\tan(n\theta)$ and $(\tan((n-1)\theta) + \tan\theta)/(1-\tan((n-1)\theta)\tan\theta)$ for a range of values of n and θ.

3.5. Write a program that requests and reads the real-valued coefficients of the equation $ax^2 + bx + c = 0$, and prints the equation and the roots of the equation. Your program should cover the three cases when the roots are real, equal and imaginary. The output should indicate which case applies, and for imaginary roots print separately the real and imaginary parts.

3.6. Write a program that reads in three integer values and tests whether they could be used to represent the sides of a triangle. Allow it also to indicate whether the triangle is a right-angled triangle.

3.7. Write a program which estimates the time of arrival in hours and minutes given values representing the distance to be travelled, the start time and the estimated average speed for the journey.

3.8. Zeller's congruence which determines the day of the week for any date in the Gregorian calendar (from 1582) up to 4903 (when it will be one day out) is

$$\text{day} = (\text{TRUNC}(2.61*m - 0.2) + d + y + y \text{ DIV } 4 + c \text{ DIV } 4 - 2*C) \text{ MOD } 7$$

where d is the day of the month, m is the number of the month, y is the year of the century and c the century number. The month number is 1 = March, 2 = April, . . ., 11 = January, 12 = February. January and February are also assumed to be in the previous year.

Write a program to read in a date as three integers and display the day of the week on which it falls. You will need to correct the number of the month and the year to satisfy the above convention.

3.9. Easter falls on the first Sunday following the first full moon on or after the vernal equinox, 21st March. The following algorithm will calculate for a given year y ($1582 < y < 4903$) the date of Easter.

cent := y DIV 100 + 1	determines the century
greg := trunc(3*cent/4) − 12	determines the Gregorian correction to allow for 'non-leap' years such as 1800 and 1900
gold := y MOD 19 + 1	determines the position of the calendar moon
c := (8*cent + 5) DIV 25 − 5 − greg	determines the clavian correction
e := 5*y div 4 − greg − 10	
moon := (11*gold + 20 + c) MOD 30	determines the age of the moon on January the first

IF (moon = 25) AND (gold > 11) OR (moon = 24) THEN moon := moon + 1
d := 44 − moon
IF d < 21 THEN d := d + 30
d := d + 7 − (d + e) mod 7
IF d < 31 THEN Easter := march d ELSE Easter := april (d−31)

Write a program to calculate the date of Easter for any given year.

3.10. Write a program to determine the number of days between two dates if each date is entered as three separate integers.

3.11 Write a program which reads from a terminal the data from a frequency distribution in the order $x_1, f_1, x_2, f_2 \ldots x_n, f_n$ (where f_1 is the frequency of x_1) and calculates the mean using the formula:

$$\bar{x} = \frac{x_1 f_1 + x_2 f_2 + \ldots + x_n f_n}{f_1 + f_2 + \ldots + f_n}$$

4 Sequences and Series

4.1 Introduction

We now have sufficient knowledge of a programming language to allow us to begin to explore the application of the computer to some simple mathematical ideas. An important aspect of computing is the ability to perform repeated operations and the Pascal language allows us the facility to do this either a pre-determined number of times or until a condition is satisfied. An area of mathematics in which one can apply this is to problems concerning sequences and series. Sequences consist of sets of numbers, each of which is generated from previous values in the sequence according to some rule. Hence, by repeated application of the rule we can generate the sequence of numbers from any starting value. In the case of series, we are usually interested in their sum which can be achieved by the repeated operation of adding the next term in the series to the current value of the sum. Thus, in computation terms, both concepts can be handled using the looping structures discussed in the previous chapter. Here we will describe and illustrate some of the problems in handling sequences and series on a computer. For example, from a mathematical viewpoint we are often interested in the convergence of the sequence or the sum to a finite limit. We therefore need to consider the problem of deciding when we have gone sufficiently far in our calculations to have achieved a suitable approximation to the limit.

The ideas covered in this chapter serve as further illustrations of using the looping constructions in Pascal. They also allow us to

(1) introduce techniques for solving mathematical problems whose solution can be approached using a sequence of values,
(2) consider ways of making some calculations easier, and in some instances possible,
(3) illustrate how numerical problems hinder the transition from a mathematical formulation to a numerical solution.

43

4.2 Sequences

4.2.1 Recurrence relationships

Consider a population of female rabbits which reproduces according to the following rules:

(1) The population starts with one female rabbit.
(2) Each female rabbit produces 1 female offspring each year starting from age 2.
(3) No female offspring dies.

Thus at year $(n+1)$ the population consists of all female rabbits that existed in year n plus the offspring (one per rabbit) of all female rabbits existing in year $(n-1)$. If we let f_n equal the number of rabbits existing in year n, then the sequence which represents the number of rabbits existing in any given year is determined by the recurrence relationship

$$f_{n+1} = f_n + f_{n-1}$$

which relates the next value in the sequence to current and past values. How can we write a program to generate such a sequence?

The calculations can be carried out using some form of repeated operation. To commence the process we will need to set starting conditions by giving values for both f_0 and f_1. It is obvious that the sequence of values generated will continue to increase; hence, to terminate the process we will need to decide in how many terms we are interested. We will also need to consider what output is required.

Suppose we use INTEGER variables *lastyear, thisyear* and *nextyear* to represent f_{n-1}, f_n, and f_{n+1} respectively. For our stopping criterion, we will use *maxf* to represent the largest value in which we are interested and produce the sequence up to this value. Again, this will be an INTEGER variable.

We can represent our problem in the following schematic way.

Step 1. Set up initial values. This can be done either by allowing them to be read into the program or assigning values within the program.
Step 2. Generate the sequence up to the maximum value printing out each term as it is generated.

This step consists of REPEATing the following substeps:

Step 2a. Calculate *nextyear := thisyear + lastyear*
Step 2b. Output the next term in the sequence, that is, *nextyear*
Step 2c. Replace *lastyear* by *thisyear* and *thisyear* by *nextyear*

UNTIL *nextyear* exceeds *maxf*.

The REPEAT loop is the most suitable choice since there is a condition to be tested for exit and we require the loop to be obeyed at least once.

Translating this into Pascal code we produce the following program:

```
PROGRAM fibonacci(INPUT,OUTPUT);

VAR   lastyear,thisyear,nextyear,maxf : INTEGER;

BEGIN

(*step 1*)
    WRITE('Type in the initial values');
    READ(lastyear,thisyear);
    WRITE('Type in the maximum value');
    READ(maxf);
    WRITELN('The sequence of yearly populations is :');
(*step 2*)
    REPEAT
(*step 2a*)
        nextyear := lastyear + thisyear;
(*step 2b*)
        WRITE(nextyear:5);
(*step 2c*)
        lastyear := thisyear;
        thisyear := nextyear;
    UNTIL nextyear >= maxf;
END.
```

The result of running this program is the following:

```
Type in the initial values 1 1
Type in the maximum value 5000
The sequence of yearly populations is:
  2    3    5    8   13   21   34   55   89  144  233  377  610  987
1597 2584 4181 6765
```

As can be seen from the output, the program as it stands produces the sequence of numbers including the one above the specified value since the full set of statements in the loop is obeyed once it is entered. To achieve the stated objective we can make the statement that displays the next value of the sequence conditional on the value of *nextyear* being less than *maxf*. That is, we can replace the output statement by the line

IF nextyear<maxf THEN WRITE(nextyear:5);

The alternative solution is to change the condition at the end of the loop to

UNTIL (lastyear+thisyear)>=maxf;

Another, and simpler, solution is to display the value of *thisyear* at the start of the repeat loop, rather than the value of *nextyear*. The effect of this would be to include the value of the initial population at the beginning of the sequence.

The sequence of numbers that this particular recurrence relationship generates is called the Fibonacci sequence after the Italian mathematician who first considered it. The numbers in such a sequence have some interesting mathematical properties, not least of which is the result that the ratio of any term to the previous term converges to the golden ratio $(\sqrt{5} + 1)/2$.

Recurrence relationships, which relate future values to past and present values, are also known as difference equations and occur in many areas of mathematical application. The generation of populations can be described by recurrence relationships since the next generation is determined by the previous ones. (Obviously, a more realistic set of assumptions would be required than in our example, but it is often surprising how good a description of the real world we can get with a set of very simple assumptions.)

In physics and applied mathematics such processes are often described using differential equations, reflecting the fact that time and space are continuous. However as we shall see later on, once we try to solve them on a computer we have to resort to recurrence relationships, using them as a discrete approximation to the differential equation.

4.2.2 Sequential mappings

A sequence is said to be generated by a sequential mapping when each succeeding value is a function of previous values. The simplest form is the case when $x_{n+1} = f(x_n)$.

For example, if $f(x) = x^2$ then the sequence 2, 4, 16, 256, . . . is that which is generated starting from $x=2$.

Clearly this sequence diverges, but as we see in the next example, we can also have sequential mappings that generate a convergent sequence. We say that a sequence of values converges to a limit if a stage can always be found such that any subsequent values differ from some limiting value by less than a pre-set chosen amount.

Thus, for example, if $f(x) = 1 + x/2$ we would, starting from $x_1=1$, generate the sequence of terms $x_1 \ldots x_n$ given by

$$x_1 = 1$$
$$x_2 = 1 + 1/2$$
$$x_3 = 1 + (1+1/2)/2 = 1 + 1/2 + 1/4$$
$$\cdot$$
$$\cdot$$
$$\cdot$$
$$x_n = 1 + 1/2 + 1/4 + \ldots + 1/2^{n-1} = (1 - (1/2)^n)/(1 - 1/2)$$

Clearly, as n tends to infinity, x_n tends to 2 and the sequence converges. Hence we can state any value such that the subsequent terms of the sequence are at some stage within that distance of the limiting value.

In such a case it can be shown that the difference between any two successive values is sufficiently small, or their ratio is sufficiently close to 1. While this is true if convergence has occurred, it can also be true in some cases where convergence is not possible, for example $\Sigma\ 1/n$. The use of these as possible stopping rules does not therefore guarantee convergence, but it will include all cases for which convergence occurs. We discuss in the next chapter the implications of using each of these rules.

In terms of programming steps, the operation of generating the above sequence can be broken down as follows:

Step 1. Specify the initial value and the accuracy required.

Step 2. Generate the sequence until the stopping rule is satisfied.

We can then describe this step in terms of the operations that are repeated:

Step 2a. Save the current value of the sequence

Step 2b. Generate the next term in the sequence

Step 2c. Check if the stopping rule is satisfied

Using *oldterm* and *nextterm* as identifiers to represent the current and the next terms in the sequence, we can translate this into the following Pascal code using a REPEAT loop to generate the sequence. If we were to execute these statements, then on exit from the loop the identifier *nextterm* would contain an approximation to the limit of the sequence.

```
epsilon := 0.0001;
nextterm := 1;
REPEAT
     oldterm := nextterm;
     nextterm := 1 + nextterm/2;
UNTIL ABS(nextterm/oldterm - 1) < epsilon;
```

The accuracy of the approximation is determined by the value given to epsilon. The smaller the value the more accurate the answer, but the longer the calculation.

From our mathematical knowledge of this sequence we know that it converges so no problem would arise in using this particular stopping rule in this case. In general, though, we are going to be using the computer in situations where we do not have prior knowledge on the convergence of the sequence. If the sequence does not converge and we use the above stopping rule we will never leave the loop and the program would run indefinitely, becoming stuck in this particular loop. It is therefore much safer, and good programming practice, to include an alternative stopping condition which we know will be reached in a finite time whether the sequence converges or not. We could, for example, count the number of iterations that are performed and stop when this reaches a stated number. The effect of this is to terminate the loop even if the sequence diverges. The alterations to the Pascal code to allow this are highlighted in italics in the following improved version of the program.

```
n:=0;
epsilon := 0.0001;
nextterm := 1;
REPEAT
     oldterm := nextterm;
     nextterm := 1 + nextterm/2;
     n := n + 1;
UNTIL (ABS(nextterm/oldterm - 1) < epsilon)
                              OR (n = 100);
```

A third, and sometimes necessary, alternative is to stop the generation of the sequence if it is diverging rapidly. This would result in a runtime error of overflow being generated since the next term in the sequence would become greater in magnitude than the largest value that the computer can accommodate. We can avoid this by testing if the value of *nextterm* has become very large. (Greater than half the largest number possible is a reasonable value for 'large'.) An alternative method would be to consider the ratio *oldterm/nextterm*. If this was very small, it would imply that the value of *nextterm* was very much larger than the value of *oldterm*. Hence the sequence of values was diverging.

Thus the final statement in the loop should be of the following form

```
UNTIL (ABS(nextterm/oldterm - 1) < epsilon OR
       (N = 100) OR (ABS(oldterm/nextterm < 1E-10);
```

Obviously, we will need to check which of these conditions has in fact been used to exit from the loop before making use of the result.

4.2.3 An application

Convergent sequences can prove a useful device if we are searching for the zero of a function. For example, if we want to find a root of

$$x^2 + 2x - 1 = 0$$

we can rewrite the equation as

$$x = 1/(2+x)$$

Then it can be shown that generating the sequence using the sequential mapping with $f(x) = 1/(2+x)$ will produce the positive root of the equation as its limit from nearly all starting values. Obvious exceptions are $x = -2$ or $-1 - \sqrt{2}$. (The first would cause an error message, indicating division by zero, while the second is the other root.)

It can also be shown that all starting values generated by

$$y_{n+1} = 1/y_n - 2 \text{ with } y_1 = -2.5$$

would also cause division by zero problems after n loops, since all these values will eventually produce the value-2 in the sequence of x values.

The sequence could be generated by the loop:

```
n:=0;
epsilon := 0.0001;
nextterm := 1;
REPEAT
    oldterm := nextterm;
    nextterm := 1 + nextterm/2;
    n := n+1;
UNTIL (ABS(nextterm/oldterm - 1) < epsilon) OR (n = 100) ;
```

We call this use of a sequential mapping an iterative process and it plays an important role in non-linear mathematics, one of the areas in which we use computers since algebraic solutions cannot usually be derived.

4.2.4 Recursive computation

Recursive methods can also be useful in helping to make some calculations easier to program. In some cases it can even help to make them possible! Consider, for example, the problem of calculating

$$\frac{n! \, e^n}{n^n \sqrt{n}}$$

For large n this is not in fact a large number, but its component parts $n!$, e^n and n^n would soon be larger than the maximum value allowed on the computer. However

$$\frac{n! \, e^n}{n^n} = \frac{ne}{n} \frac{(n-1)! \, e^{n-1}}{n^{n-1}} = \frac{ne}{n} \frac{(n-1)e}{n} \frac{(n-2)! \, e^{n-2}}{n^{n-2}} = \cdots$$

hence it can be generated by a sequence of steps using the sequential mapping

$$x_k = k \, e \, x_{k-1}/n, \text{ starting from } x_0 = 1$$

The following program illustrates its use to calculate the function for a particular value of n.

```
PROGRAM stirlingsapproximation (INPUT,OUTPUT);
CONST e = 2.718281828;
VAR n,j : INTEGER;
    x : REAL;
BEGIN
(*input stage*)
    WRITELN(' Type in the value of n');
    READ(n);
(*calculation stage*)
    x := 1;
    FOR j := 1 TO n DO
        x := x * j * e / n;
(*output stage*)
    WRITELN('answer = ',x / SQRT(n));
END.
```

If we ran such a program, we would find that the answer for large n tended towards the value of $\sqrt{(2\pi)}$. It does, in fact, illustrate Stirling's approximation which states that for large n

$$n! \approx \sqrt{(2\pi)}\, e^{-n}\, n^{n+1/2}$$

in the sense that the ratio of the left-hand side to the right-hand side tends towards 1.

Thus, the right-hand side can be used as an approximation to $n!$ for large n even though the difference between the two sides increases rapidly with n. The percentage difference is in fact tending towards zero, reflecting the fact that the number of significant figures in agreement is growing. Thus, in computational terms, because of the way real numbers are approximated, the right-hand side can be used as a computationally efficient approximation to $n!$.

4.3 Series

Another area in which we can use repeated operations is the summing of series. Suppose, for example, we wanted to calculate

$$\sum_{n=0}^{\infty} 1/n^3 = 1 + 1/2^3 + 1/3^3 + \dots$$

whose exact value is unknown.

Obviously, we cannot include an infinite number of terms; we therefore need to devise some rule for stopping the operation. Various possibilities are available for determining how many terms to include in a series. For example, we could stop when:

(1) the number of terms reached a pre-determined number, or
(2) the size of the next term to be added is sufficiently small, or
(3) the value of the result can be shown by theory to be acceptably close to the true value.

If we use the first we could adequately program the operation using a **FOR** loop as follows:

```
sum := 0;
FOR i := 1 TO n DO
   BEGIN
      reciprocal := 1/i;
      sum := sum + reciprocal * reciprocal * reciprocal
   END;
```

Note that the calculation of the term $1/i^3$ using the formula as written may cause problems for large values of i. The value of i^3 may become greater than the largest integer allowed on the computer, which will almost certainly result in problems with the calculation of the sum. To avoid this, we calculate the value of $1/i$ and obtain the third power of this expression by repeated multiplication.

In this example, the existence of the sum to infinity can be established using mathematical methods. Furthermore, if the last term included in the sum is the nth, then the error (called the truncation error) is equal to that part of the sum omitted. That is

error = Σ $1/i^3$ where the summation is over the range $n+1$ to ∞

This is approximately equal to

$$\int_{n+1}^{\infty} (1/x^3)\, dx = 1 / [2(n + 1)^2]$$

Thus, in this case, we can write an algorithm which enables us to calculate an estimate of the sum to a sufficiently small error. It poses us the problem of how to interpret the phrase 'sufficiently small'. For example, in the following we stop if the next term is smaller than 0.00001, but this in fact only gives us the sum accurate to three decimal places.

It should be noted that we cannot use the computer to establish the convergence of a series but, provided that the convergence can be established mathematically, we can always calculate the sum to within some calculable error.

The steps enabling us to calculate a sum in this way are:

Step 1. Set the value of the sum to zero.
Step 2. Set the term number to zero.
Step 3. Calculate the sum.
     Step 3a. Increase the term number by 1.
     Step 3b. Calculate the next term in the series.
     Step 3c. Add the value of the next term to the sum.

If we calculate the sum until the next term adds a negligible amount to the sum, or a total of n terms have been included, then the natural way to program this is using a REPEAT loop, since we are usually going to include at least one term, although we could also use the alternative program shown using a WHILE loop.

```
epsilon := 0.00001;
sum := 0;
i := 0;
REPEAT
   i := i + 1;
   reciprocal := 1/i;
   term := reciprocal * reciprocal * reciprocal;
   sum := sum + term;
UNTIL (term < epsilon) OR (i = n) OR (sum > 10E38);
```

```
sum := 0;
epsilon := 1E-5;
i := 0;
term := 1;
WHILE (term > epsilon) OR (i < n) OR (sum < 10E38) DO
  BEGIN
    i := i + 1;
    reciprocal := 1 / i;
    term := reciprocal * reciprocal * reciprocal;
    sum := sum + term;
  END;
```

Note that in the WHILE example the quantities used in the condition (in this case the identifiers *term, i* and *n*) must all be given values before we reach the loop.

4.3.1 Calculation of exp (x) using the exponential series

We conclude this chapter by considering the calculation of e^x which, while it can be done using the supplied function EXP(x) described in section 2.4, provides a useful illustration of the calculation of series and some of the numerical problems that may be encountered. It also serves as an example to illustrate the balance between computational efficiency and numerical accuracy that should always be considered when developing any program.

The Maclaurin expansion of $\exp(x)$ is

$$1 + x + x^2/2! + x^3/3! + \ldots$$

We know that this converges for all values of x, hence we could obtain an adequate approximation by terminating the summation once the next term was sufficiently small. The effect of this is to approximate the infinite sum by

$$\sum_{i=0}^{n} \frac{x^i}{i!}$$

The truncation error is therefore the difference between the true value and the approximation and is

$$\frac{x^{n+1}}{(n+1)!} + \frac{x^{n+2}}{(n+2)!} + \ldots$$

$$= \frac{x^{n+1}}{(n+1)!} \left\{ 1 + \frac{x}{n+2} + \frac{x^2}{(n+2)(n+3)} + \ldots \right\}$$

$$< \left\{ \frac{x^{n+1}}{(n+1)!} + \frac{x^{n+2}}{(n+2)!} \right\} \left\{ 1 + \frac{x^2}{n^2} + \frac{x^4}{n^4} + \ldots \right\}$$

$$= \frac{x^{n+1}}{(n+1)!} \left\{ 1 + \frac{x}{(n+2)} \right\} \left\{ 1 - \left(\frac{x}{n} \right)^2 \right\}^{-1} \quad \text{for all } x$$

From this we can see that the error, as well as being dependent on n, is also determined by x. This suggests that more terms of the series will be required to obtain the same accuracy as the value of x increases in value.

Returning to the calculation, the steps to be followed are the same as those considered in the previous example, except that we use the fact that the ith term in the series is the previous term multiplied by x/i. (Using this fact is more efficient than calculating $x^i/i!$ for each i. It also avoids the potential overflow problem discussed in section 4.2.3.)

We could therefore calculate it using the following code:

```
i := 1;
sum := 1;
term := 1;
REPEAT
    i := i + 1;
    term := term * x / i;
    sum := sum + term;
UNTIL ABS(term) < epsilon;
```

where epsilon is some pre-set value reflecting the accuracy that we wanted to achieve.

Alternatively, we could perform the summation over a fixed number of terms as follows:

```
sum := 1;
term := 1;
FOR i := 1 TO n DO
   BEGIN
       term := term * x / i;
       sum := sum + term;
   END;
```

Obviously as x increases, to produce the values of e^x to the same order of accuracy as for small values of x, more terms need to be included in the series. Thus the first solution would produce the same order of accuracy for each positive value of x, but would use more terms for the larger values of x. The second solution, on the other hand, would make use of the same amount of computing time for each value of x, but produce very inaccurate results for large x. Large negative values, however, will always produce poor results using either method because of the errors that result from subtracting almost equal values (a problem which we will describe in more detail in the next chapter). Thus neither solution to the problem is completely adequate over the full range of values of x, since the first would be very expensive for some values of x, while the other would not be accurate enough for large positive values of x. Finally, neither method is capable of achieving accurate answers for large negative values.

Table 4.1	Comparison of methods for calculating $\exp(x)$

x	EXP(x)	Approximations				
		Method 1	n	% error	Method 2 (using n=12)	% error
−5.0	0.006738	0.006746	21	−1.1e−01	3.555184	−5.3e+04
−4.5	0.011109	0.011126	19	−1.6e−01	1.435443	−1.3e+04
−4.0	0.018316	0.018307	18	4.8e−02	0.530159	−2.8e+03
−3.5	0.030197	0.030202	17	−1.4e−02	0.190022	−5.3e+02
−3.0	0.049787	0.049796	15	−1.9e−02	0.091295	−8.3e+01
−2.5	0.082085	0.082081	14	4.5e−03	0.090462	−1.0e+01
−2.0	0.135335	0.135328	12	5.5e−03	0.136508	−8.7e−01
−1.5	0.223130	0.223132	11	−8.6e−04	0.223222	−4.1e−02
1.0	0.367879	0.367882	9	−6.8e−04	0.367882	−6.8e−04
0.5	0.606531	0.606532	7	−2.4e−04	0.606531	−8.4e−07
0.0	1.000000	1.000000	2	0.0e+00	1.000000	0.0e+00
0.5	1.648721	1.648720	7	1.0e−04	1.648721	2.8e−07
1.0	2.718282	2.718279	9	1.1e−04	2.718279	1.1e−04
1.5	4.481689	4.481687	11	5.5e−05	4.481565	2.8e−03
2.0	7.389056	7.389046	12	1.4e−04	7.387302	2.4e−02
2.5	12.182494	12.182489	14	4.2e−05	12.168603	1.1e−01
3.0	20.085537	20.085523	15	6.7e−05	20.009152	3.8e−01
3.5	33.115452	33.115446	17	1.9e−05	32.788481	9.9e−01
·4.0	54.598150	54.598136	18	2.5e−05	53.431746	2.1e+00
4.5	90.017131	90.017104	19	3.0e−05	86.393284	4.0e+00
5.0	148.413159	148.413147	21	7.9e−06	138.307168	6.8e+00

Table 4.1 displays the results for x in the range −5 to 5 using the two methods, compared with the result from the supplied function EXP(x). The percentage error gives an indication of the accuracy that is achieved.

How can we avoid these problems? One possible method is suggested by the examination of the above results and a fairly elementary piece of mathematics. Within the range −0.5 to 0.5 the difference in accuracy or computing time can be considered negligible. Any number outside this range can be reduced to a number within this range by repeated division by 2. We also know that

$$e^x = [e^{x/2}]^2$$

Thus, repeated division of the exponent by 2 can be recovered by squaring the result a number of times equal to the number of divisions that were performed. We could therefore modify our algorithm into the following steps:

1. Reduce the number to one lying between −0.5 and 0.5 by repeated halving, keeping a count of the number of times we are required to do this.

2. Use the Maclaurin expansion to evaluate the exponential of the resulting value using one of the two stopping conditions discussed earlier.
3. Square up the result the appropriate number of times to produce the exponential of the required value.

Each of these major steps could be subdivided into minor steps before writing the Pascal code, as we have detailed in earlier examples.

The resulting code is

```
(*reduction to the range -0.5 to 0.5
    the number of divisions is accumulated in k *)
k := 0;
WHILE ABS(x) > 0.5 DO
    BEGIN
        x := x / 2;
        k := k + 1;
    END;

(*evaluation of the Maclaurin series *)
n := 8;
sum := 1;
term := 1;
FOR i := 1 TO n DO
    BEGIN
        term := term * x / i;
        sum := sum + term;
    END;

(*square up k times to produce the required answer *)
WHILE k > 0 DO
    BEGIN
        sum := sum * sum;
        k := k - 1;
    END;
```

WHILE loops are used in the first and last parts since, if the number already lies in the range −0.5 to 0.5, we will only need to use the middle part of the program and want to omit the sets of statements in both WHILE loops. For the middle block we could have used the alternative stopping criterion, but it makes little difference in practice.

4.4 Exercises

4.1. Write a program that reads in a real value s ($\geqslant 1.5$) and finds the smallest integer n such that

$$1 + 1/2 + 1/3 + \ldots + 1/n > s$$

4.2. Write a program to calculate and display the value of

$$1 + 1/2 + 1/3 + \ldots + 1/n - \ln(n)$$

for $n = 10^p$, using p values 1,2,3,4,5. Comment on your results.

4.3. Calculate the sum $1 + x + x^2 + x^3 + \ldots + x^n$ until $x^{n+1} \leqslant 0.00001$ and compare this with the answer obtained by the algebraic formula

$$S = (1 - x^{n+1})/(1 - x)$$

Also, calculate the sum for $n = 1$ to 50 when $x = 0.9$ and $x = -0.9$ and comment on the results.

4.4. Write a program to generate a sequence of numbers $P(n)$ using the recurrence relationship

$$P(n+1) = 2*P(n) + P(n-1) \quad n = 0, 1, \ldots$$

with starting values $P(0) = 1$ and $P(1) = 3$.

Use the same relationship to calculate the values of $Q(n)$ starting with $Q(0) = 1$ and $Q(1) = 2$. Calculate the ratio $P(n)/Q(n)$ for $n = 1 \ldots 20$, comparing each ratio with its square.

4.5. We have, by rearranging the equation $x^2 - 2x + 1 = 0$ as $x = 1/(x + 2)$, obtained a sequential mapping which gave us the positive root for nearly all starting values.

By using an alternative rearrangement of the equation, find a sequential mapping which will give the negative root. For what starting values will it produce a solution?

4.6. By considering a range of values for a, can you suggest a range of values for which the sequence generated by

$$x_{n+1} = a^x n$$

converges and determine its limit?

4.7. The binomial coefficients are given by

$$_nC_r = n!/((n-r)! \, r!) = n(n-1) \ldots (n-r+1)/r!$$

Write a program to print all the values of the binomial coefficients for $n = 6$ and $n = 14$.

4.8. The probability function for the binomial distribution is given by

$$\Pr(X = x) = {_nC_r} \, p^x (1-p)^{n-x} \text{ for } x = 0, 1 \ldots n \quad 0 \leqslant p \leqslant 1$$

Write a program to calculate these probabilities for $p=0.2$, $p=0.5$ and $p=0.8$ for $n=6$ and $n=14$.

4.9. The probability function for the Poisson distribution is given by

$$\Pr(X = x) = (\lambda^x/x!) \, e^\lambda \text{ for } x = 0, 1, 2 \ldots \quad \lambda > 0$$

For $\lambda = 0.95$ and $\lambda = 5$, calculate and display the probabilities that are greater than 0.001.

4.10. Write a program to estimate the values of $\sin(x)$ using the truncated Maclaurin series

$$\sum_{k=0}^{n} (-1)^k x^{2k+1}/(2k+1)! = x - x^3/3! + x^5/5! + \ldots$$

for x, taking values 0 to 20 in steps of size 5 and
for $n = 5, 10, 15, 20$ and 25.

Also estimate $\sin(x)$ by including terms in the series until the next term is sufficiently small. Record the number of terms included in the estimate.

Compare both sets of answers with that obtained using the Pascal function $\sin(x)$ and comment on your results. Suggest and program a more efficient way of evaluating $\sin(x)$ using the Maclaurin expansion.

4.11. Repeat exercise 4.7 using $\cos(x)$ and the Maclaurin series

$$\sum_{k=0}^{n} (-1)^k x^{2k}/(2k!) = 1 - x^2/2! + x^4/4! - x^6/6! + \ldots$$

5 Numerical Errors and Accuracy

5.1 Introduction

Results obtained from numerical calculations on a computer are subject to two sources of possible error. One of these is the unavoidable error which results from the way that numbers are stored on a computer, while the other error is one deliberately incurred by the numerical method we have chosen.

We have seen that when a real number is stored on a computer, only so many significant figures can be retained. The result is that a small amount of information about the number is lost because the number has been rounded. The number of significant figures retained varies from one computer to another, but is usually in the range of 7 to 14 digits. An immediate consequence of this is that arithmetic carried out on real values using a computer does not match exactly the arithmetic we usually do.

The second source of error occurs in the application of mathematical theory and results from the truncation which must occur whenever we perform any operation that in theory approaches a limit. We have already met this in the context of summing series, where we have to truncate the series when obtaining an estimate of its limit. The accuracy which can be obtained is limited by the amount of time taken to reach the limit and the numerical accuracy of the computer itself.

In this chapter we examine in more detail the way in which numbers are represented, and the effect this has on the basic arithmetic operations.

5.2 Representation of Real Numbers

A real number is represented on a computer to n-digit accuracy. Thus, a real number c is stored in exponent form $d*10^m$ where d is a decimal number between -1 and 1 having n decimal places, the first of which is non-zero, while the exponent is a two-digit integer lying between $-L$ and M. This type of representation is called a floating point representation and ensures that all real numbers are stored with the same number of significant figures. The values n, L and M are computer dependent but in all cases restrict the infinite set of real numbers to a large but finite set. Any number having more than n digits is rounded to n digits, while if the exponent m goes outside the specified range an error occurs. If $m > M$ then the error is called an overflow error while if $m < -L$ the error is called an underflow error. Occurrences of such errors usually cause the program to terminate with an appropriate error message.

However, with some compilers, when an underflow occurs the stored number is replaced by zero and the computation allowed to proceed, usually after the display of an appropriate warning, and may produce the correct numerical result. This can be a little disconcerting and we should perhaps not be too happy about such a 'successful' run since we are not in complete control of the computations being carried out.

One common possible cause of an overflow error is division by a very small number.

In representing a number c in this way it is approximated by the value c_0 such that

$$c_0 = c + \epsilon$$

where ϵ represents the error resulting from the computer representation. The *absolute error* is defined as the absolute difference between the number c and its approximation c_0.

The absolute error is therefore

$$|c - c_0| = |\epsilon|$$

For example, if we work to three-digit accuracy the number $0.437*10^3$ represents all the real numbers between 436.5 and 437.5 hence the absolute error resulting from rounding these numbers is less than $0.5*10^0$. Similarly, $0.236*10^{-3}$ represents all the real numbers between 0.0002355 and 0.0002365 hence the absolute error is $0.5*10^{-6}$. In general, if a number is represented in the form $d*10^m$, then the absolute error resulting from that representation is less than $0.5*10^{m-n}$ where n is the number of significant digits that the computer retains.

Thus, the absolute error between two numbers indicates the number of decimal places to which the numbers agree. This can, however, be a little misleading when we consider the accuracy with which two numbers of differing orders of magnitudes are stored. Thus, on a three-digit machine 436 is recorded to zero decimal places while 0.000236 is recorded to six decimal places. However, the numbers are recorded to an equal degree of accuracy, namely three significant figures. In the case of large numbers the number of decimal places is not important, while for small numbers the leading zeros do not supply any information as to the accuracy with which the number is recorded. We therefore introduce a second measurement of error that reflects the number of significant figures in a number that can be considered accurate.

The *relative error* between two numbers is the ratio of the absolute error to the absolute value of the true number. Thus, the maximum relative error

$$r = \frac{|c_0 - c|}{|c|} \leqslant \frac{|\text{maximum error}|}{|d| * 10^m}$$

$$= \frac{0.5 * 10^{m-n}}{|d| * 10^m} = \frac{0.5 * 10^{-n}}{|d|} \leqslant 0.5 * 10^{1-n}$$

since $|d| > 1/10$, because the first decimal place must be a non-zero digit, hence the relative error indicates the number of significant digits which is the same in the true number and its approximation.

5.3 Computer Arithmetic

The consequences of this method of representing numbers is seen if we examine the basic arithmetic operations. Suppose we have two numbers c_1 and c_2 which are represented on the computer by the rounded values c_{01} and c_{02} respectively. Then

$$c_{01} = c_1 + \epsilon_1 \text{ and } c_{02} = c_2 + \epsilon_2$$

where ϵ_1 and ϵ_2 are the errors in the representation.

Then $c_{01} + c_{02} = c_1 + c_2 + \epsilon_1 + \epsilon_2$

and $c_{01} - c_{02} = c_1 - c_2 + \epsilon_1 - \epsilon_2$

Thus the absolute errors are

$$|c_1 + c_2 - c_{01} - c_{02}| = |\epsilon_1 + \epsilon_2| \quad \text{and}$$

$$|c_1 - c_2 - c_{01} + c_{02}| = |\epsilon_1 - \epsilon_2| \quad \text{respectively.}$$

Hence since $\qquad\qquad |\epsilon_1 + \epsilon_2| \leqslant |\epsilon_1| + |\epsilon_2|$

and $\qquad\qquad\qquad |\epsilon_1 - \epsilon_2| \leqslant |\epsilon_1| + |\epsilon_2|$

the absolute error that results from both addition and subtraction is less than the sum of the absolute errors in the numbers involved.

If we consider multiplication then

$$c_{01} * c_{02} = c_1 * c_2 + c_1 * \epsilon_2 + c_2 * \epsilon_1 + \epsilon_1 * \epsilon_2$$

Because the errors are small, the term $\epsilon_1 * \epsilon_2$ can be discounted since it is an order of magnitude smaller than the other two error terms. Hence the absolute error is approximately

$$|c_1 * c_2 - c_{01} * c_{02}| \approx |c_1 * \epsilon_2 + c_2 * \epsilon_1|$$

Hence the relative error is

$$\frac{|c_1 * c_2 - c_{01} * c_{02}|}{|c_1 * c_2|} \approx \frac{|\epsilon_1|}{|c_1|} + \frac{|\epsilon_2|}{|c_2|} \leqslant \frac{|\epsilon_1|}{|c_1|} + \frac{|\epsilon_2|}{|c_2|}$$

That is, the relative error in the multiplication is less than the sum of the relative errors in the two original numbers.

Finally we consider division.

$$\frac{c_{01}}{c_{02}} = \frac{c_1 + \epsilon_1}{c_2 + \epsilon_2} = \frac{c_1(1 + \epsilon_1/c_1)}{c_2(1 + \epsilon_2/c_2)}$$

which is approximately

$$\frac{c_1}{c_2} \left\{ 1 + \frac{\epsilon_1}{c_1} \right\} \left\{ 1 - \frac{\epsilon_2}{c_2} \right\}$$

$$= \frac{c_1}{c_2} \left\{ 1 + \frac{\epsilon_1}{c_1} - \frac{\epsilon_2}{c_2} \right\}$$

ignoring the terms in $\epsilon_1 * \epsilon_2$. Hence, the relative error is approximately

$$\left| \frac{\epsilon_1}{c_1} - \frac{\epsilon_2}{c_2} \right| \leqslant \frac{|\epsilon_1|}{|c_1|} + \frac{|\epsilon_2|}{|c_2|}$$

Thus the maximum relative error is less than the sum of the relative errors.

Hence the rules of arithmetic are such that for addition and subtraction the resulting accuracy is expressed in terms of the number of decimal places, while for multiplication and division the resulting accuracy is expressed in the number of significant figures.

It is clear from this that if we perform a series of arithmetic operations then the magnitude of the maximum possible error increases with each operation and hence it is possible for large errors to occur. Fortunately, while the original error is equally likely to have occurred anywhere in the range covered by the error, the extremes are less likely to have occurred in the accumulation. There are, however, a few cases over which particular care needs to be taken, and we illustrate these by example.

5.3.1 Examples

If we have a sequence of arithmetic operations to perform, we must be aware that at each stage of the sequence the current result is rounded. Hence in some cases both the order of the operation and the relative magnitudes of the numbers involved can have an effect.

Suppose we have a machine that retains real numbers to three significant numbers only.

We begin by considering the results obtained by applying the four basic arithmetic operations to the numbers $x = 6/7$ and $y = 1/6$, stored as 0.857 and 0.167 respectively on a three-digit machine. Table 5.1 summarises the results. From this we see that all the calculations have a relative error of less than 0.00372 in magnitude, and hence produce satisfactory three-digit results.

Table 5.1

Operation	Correct value	Approximate value	Absolute error	Relative error
$x + y$	1.0238 (43/42)	1.02	0.0038	0.00372
$x - y$	0.6905 (29/42)	0.690	0.0005	0.00069
$x * y$	0.1429 (1/7)	0.143	0.0001	0.00100
x / y	5.1429 (36/7)	5.13	0.0129	0.00250

Suppose we now consider each of the operations in turn and explore cases which are likely to cause problems.

Consider the result of adding the numbers 174, 842 and 356.

The first step adds 174 to 842 which gives 1016 which is rounded to $0.102*10^4$ when it is stored — that is, 1020.

This is then added to 356, giving 1376 which is stored as $0.138*10^4$ — that is, 1380. Thus, in terms of the arithmetic operations carried out on a computer working to three significant figures, we get 1380 as the answer compared with the correct answer 1372. The absolute error is 8, while the relative error is 0.0058.

However, if we had added the numbers in the order 174, 356 and 842 we would proceed as follows.

174+356 = 530 which is stored as $0.530*10^3$ (that is, no information is lost)

Then 530+842 = 1372 which is stored as $0.137*10^2$, giving us the more accurate answer 1370. The absolute error in this case is 2, while the relative error is 0.0015.

Thus, in certain cases when repeated operations are being carried out we can sometimes improve the accuracy by adding the small numbers first before the larger ones. This is particularly so when calculating the sum of a series of numbers that converges. The later terms in the sequence will be much smaller than the earlier terms. Hence, if we sum the sequence in the conventional way from left to right, the contribution of the later terms will be ignored since they will not contribute to the number of significant numbers in the sum. However, if we add from right to left then the sum of the first few terms will be the same order of magnitude as these very small terms. Hence they will make a significant contribution to the sum which will continue to be present once the first term in the sum is included. Thus, adding in the conventional way will result in a small numerical error which can be reduced by summing from right to left.

This possible problem resulting from adding numbers is highlighted if we add three numbers which differ in order of magnitude. Suppose we add 684.3, 13.42 and 0.1652. On our three-digit machine these are represented as 684, 13.4 and 0.165. Suppose we add these numbers in the order shown. Then

684 + 13.4 gives 697.4   which is rounded to 697
697 + 0.165 gives 697.165 which is rounded to 697

The true answer is 697.8852, thus the absolute error is 0.8852 and the relative error is 0.00127. Notice here that the smallest value has no effect on the result which, while having a large absolute error, has a small relative error and agrees with the correct result as far as is possible when working to three digits.

However, if we had added them in reverse order then

13.4 + 0.165 gives 13.565 which is rounded to 13.6
13.6 + 684   gives 697.6   which is rounded to 698

This has an absolute error of 0.1148 and a relative error of 0.000164. Hence, this clearly illustrates the numerical errors involved and also suggests that adding from small to large rather than large to small will help to overcome the problem.

With subtraction, the main problem arises when the two numbers are very similar in order of magnitude. For example, if we have a three-digit machine on which we calculate $13\,743 - 13\,621$. The numbers are stored as $0.137*10^5$ and $0.136*10^5$ and the result of the subtraction is $0.001*10^3$ (that is 100) which compares with correct answer of 122, and results in a 17 per cent error. The addition results in $0.273*10^5$ — that is, 27 300 — compared with the correct answer of 27 364 — a 0.2 per cent error.

Subtraction of approximately equal quantities is called *cancellation*, since the accurate leading digits in the numbers cancel one another, leaving us with the digits that are more prone to rounding error. We have already met one example where this arises, the calculation of $\exp(x)$ when x is a large negative number.

Table 5.2 illustrates the result of further operations applied to the result of such a subtraction using $y = 1/6$, $u = 0.1652$, $w = 1/3000$ and $v = 684.3$. These numbers are stored as 0.167, 0.165, 0.000333 and 684 respectively.

Table 5.2

Operation	Correct value	Approximate value	Absolute error	Relative error
$y - u$	0.00147	0.00200	0.000533	0.3636
$(y-u)/w$	4.40000	6.01000	1.61	0.3659
$(y-u)*v$	1.0036	1.37	0.3664	0.3650

We see that the large relative error produced as a result of the subtraction is unaffected by the subsequent multiplication and division, but the absolute error is very much magnified when we divide by a relatively small number or equivalently when we multiply by a relatively large number.

In the case of multiplication the effect of a sequence of operations can be visualised by considering the case of multiplying numbers on a computer which works to n digits. In this case, if two n-digit numbers are multiplied together then the result is a $2n$-digit or a $2n-1$-digit number which must be rounded to n significant figures before the next operation is carried out.

Thus, if we again consider calculations working to three digits, the result of multiplying 234*12*476 is as follows:

$$234*12 = 2808 \text{ which is stored as } 0.281*10^4 \text{ (that is, 2810)}$$

then

$$2810*476 = 1\,337\,560 \text{ which is stored as } 0.134*10^7 \text{ (that is, } 1\,340\,000)$$

which compares with the correct answer, 1 336 608.

In the case of division, it is easier to see the result of computer division since the computer cannot handle fractions. 1/3, for example, is represented as 0.333 on a three-digit machine which is slightly under a third. Hence, as a result of losing the ability to manipulate fractions, we lose a degree of numerical accuracy.

To illustrate the handling of numerical errors in the context of a problem, we consider the calculation of the roots of the quadratic equation $x^2 - 11x + 1 = 0$ on a three-digit machine. Using the standard formula the roots are given by

$$\frac{-b \pm \sqrt{(b^2 - 4ac)}}{2a} \quad \text{where } b = -11 \text{ and } a = c = 1$$

On a three-digit machine, the value of $b^2 - 4ac$ is correctly stored as 117. However, the square root is 10.8167 which is stored as 10.8. Using this value, the larger root is calculated as $(11+10.8)/2 = 10.9$ while the smaller is $(11-10.8)/2 = 0.100$ whereas the correct values are 10.908 and 0.091674 respectively. Thus, using a three-digit machine the relative errors are 0.0008 for the larger root and 0.091 for the smaller root. Clearly the latter is not satisfactory. If, however, we had calculated the smaller root using $c/$(larger root), then we get 0.0917 which has a relative error of 0.0003. The reason for the improvement using this second method is the fact that it avoids the subtraction of two numbers which are of the same order of magnitude.

5.3.2 Comments

It should be stressed that most computers work to somewhere between 7 and 14 digits, hence the problem is not as serious as suggested by the examples described using three-digit calculations.

Furthermore, while the absolute error for a single number is equally likely to have a value anywhere up to the maximum value, the distribution of the error resulting from any of the arithmetic operations is not as bad. The distribution of the errors resulting from any of the basic operations is a triangular distribution with errors near zero being more likely than values near its maximum. This is because of the way in which positive and negative errors combine. Thus, while there is a possibility of the absolute error being twice the maximum absolute error in each of the numbers, it is not as likely to occur as a much smaller value. (*Note:* once we have covered random number generation in chapter 15 we can illustrate this by simulation. See exercise **15.4**.)

Hence there is an element of doubt about the accuracy of the last one or two significant digits of any numerical values produced on a computer. However, the first few digits which usually are those in which we are interested, can, in most cases, be used with some confidence. Some problems are particularly prone to numerical error and these are discussed at the appropriate stage, but in general we can proceed safely provided that we are aware of the potential problems and prepared to adapt algorithms to handle them.

5.4 Order of Calculation

One method of adapting algorithms to avoid potential numerical problems is to re-order the way in which the calculations are carried out. Thus, for example, if the natural order in which a problem is posed results in the subtraction of two very similar values, it may be possible to re-order the operations so that the subtraction, when it occurs, does not involve similar values.

The re-ordering of operations to improve numerical accuracy can be clearly seen if we consider the problem of calculating a sum of the form

$$S = 1 + 1/2^4 + 1/3^4 + \ldots + 1/n^4 + \ldots$$

As n becomes larger the next term, $1/n^4$, to be added to the sum becomes smaller, until eventually it is preceded by as many zeros after the decimal place as the number of digits to which the computer works. Hence adding the next, and any subsequent, terms to the sum will result in no further contribution being made. However, if we had added the small terms first, it is quite likely that their sum will affect the last few digits of the overall sum. Hence, by adding the series from right to left — that is, from the smallest term to the largest term — we will obtain a more accurate answer, since the accumulation of the small terms will be allowed for in the sum.

A second example, which illustrates how errors accumulate, is the following. Consider the integral

$$I_n = \int_0^1 x^n \, e^{(x-1)} \mathrm{d}x$$

Then $I_0 = (1 - e^{-1})$

and $I_n = 1 - n \, I_{n-1}$

If we use this as an algorithm to generate the sequence of integrals, it leads to unstable answers since the rounding errors grow and dominate the sequence. If we let $\epsilon_n = I_n - \tilde{I}_n$, where $\tilde{I}_n$ is the value obtained on a computer and I_n is the true value, then

$$\epsilon_n = I_n - \tilde{I}_n = 1 - n\,I_{n-1} - 1 + n\,\tilde{I}_{n-1}$$

$$= (-1)\,n\,\epsilon_{n-1}$$

$$= (-1)^n\,n!\,\epsilon_0$$

Thus $\epsilon_1 = -\epsilon_0$

$\epsilon_2 = 2\epsilon_0$

$\epsilon_3 = -6\epsilon_0$, etc.

Hence the error grows very rapidly and soon the value of the sequence I_n will produce an overflow error.

However $\quad I_n - I_{n-1} = \displaystyle\int_0^1 (x^n - x^{n-1})\exp(x-1)\,dx$

and since $0 < x < 1$, then $x^n < x^{n-1}$

hence $\quad\quad I_n - I_{n-1} < 0$

That is

$$I_n < I_{n-1}$$

Hence as n increases I_n decreases, and since the integral is always positive it must tend to zero as n tends to infinity. Thus, to obtain estimates of the integral for values of n, if we start with $I_n = 0$ for some large n, then using the sequential mapping in reverse, that is

$$I_{n-1} = (1 - I_n)/n \text{ for } i = n, n-1, n-2 \ldots$$

we produce a stable sequence in which the errors do not explode. Thus, for this example, using a backward recurrence is more stable than using a forward recurrence. The results obtained by using both recursions are shown in table 5.3.

Table 5.3   Calculation of $\int_0^1 x^n e^{x-1}\, dx$ using backward and forward recursions

Forward recursion		Backward recursion	
n	$I(n)$	n	$I(n)$
0	6.32121e−01	20	0.00000e+00
1	3.67879e−01	19	5.00000e−02
2	2.64241e−01	18	5.00000e−02
3	2.07277e−01	17	5.27778e−02
4	1.70893e−01	16	5.57190e−02
5	1.45533e−01	15	5.90176e−02
6	1.26802e−01	14	6.27322e−02
7	1.12384e−01	13	6.69477e−02
8	1.00929e−01	12	7.17733e−02
9	9.16375e−02	11	7.73522e−02
10	8.36254e−02	10	8.38771e−02
11	8.01209e−02	9	9.16123e−02
12	3.85497e−02	8	1.00932e−01
13	4.98854e−01	7	1.12384e−01
14	−5.98396e+00	6	1.26802e−01
15	9.07594e+01	5	1.45533e−01
16	−1.45115e+03	4	1.70893e−01
17	2.46706e+04	3	2.07277e−01
18	−4.44069e+05	2	2.64241e−01
19	8.43731e+06	1	3.67879e−01
20	−1.68746e+08	0	6.32121e−01

5.5   Exercises

5.1. Write a program to calculate the values of

$$a = \sqrt{(n+1)} - \sqrt{n} \qquad b = 1/(\sqrt{(n+1)} + \sqrt{n})$$
$$c = \sqrt{n}(\sqrt{(1 + 1/n)} - 1) \quad d = 1/(2\sqrt{n})$$

for $n = 10^p$ for $p = 1,2,3 \ldots 20$.

Observing that the formulae are alternative ways of representing the same value, comment on what you observe as output from your program.

5.2. The equation $x^2 - bx + 1 = 0$ has two roots, x_1 and x_2, whose product $x_1 x_2 = 1$. Write a program to calculate x_1, x_2 and x_3 from the formulae

$$x_1 = (-b + \sqrt{(b^2 - 4ac)})/2a$$
$$x_2 = (-b - \sqrt{(b^2 - 4ac)})/2a$$
$$x_3 = 1/x_2$$

Comment on the observed values when b takes values

$$b = (10 + 10^{-n})\, 10^p \text{ for } n = 3, 6, 9, 12$$
$$\text{and } p = 0, 3, 6, 9, 12$$

5.3. Calculate the values generated by the sequential mapping

$$U_{n+1} = (x+1)\, U_n - 1 \text{ starting from } U_0 = 1/x. \text{ For } x = 2, 3, 4, 5 \dots$$

by hand and on the computer. Comment on and explain what you observe.

5.4. Calculate the sum $S = 1 + 1/2^4 + 1/3^4 + \dots + 1/N^4$ for $N = 10^p$ using $p = 2, 3, 4 \dots$ by adding the terms from left to right and right to left. Examine the results and comment on any differences between them.

Compute $\mathrm{SQRT}(\mathrm{SQRT}(90*S))$ and guess the value of the infinite sum of inverse fourth powers.

5.5. The sum of chords $AA_1 A_2 \dots A_n$ which are the half-perimeter of a regular $2n$-sided polygon inscribed in a circle of radius 1 is

$$l_n = 2n \sin(\pi/2n)$$

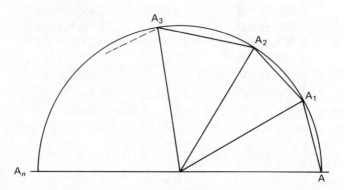

It is obvious that $l_n \to \pi$ as $n \to \infty$.
For $x \in [0, \pi/2]$

$$\sin(x/2) = \sqrt{(0.5\,(1 - \sqrt{(1-\sin^2 x)}))}$$

Hence

$$l_{2n} = 4n\sqrt{(0.5\,(1-\sqrt{(1-(l_n/2n)^2)}))}$$

taking $n = 2^{k-1}$ and $a_k = l_n$

we have

$$a_1 = 2, \quad a_{k+1} = 2^{k+1} \sqrt{(0.5(1 - \sqrt{(1 - (a_k/2^k)^2)}))}$$

Further, for $x \neq -1$

$$1 - x = (1 - x^2)/(1 + x)$$

Thus, if we let $x = \sqrt{(1 - (a_k/2^k)^2)}$, we also have

$$a_1 = 2, \quad a_{k+1} = a_k \sqrt{(2/(1 + \sqrt{(1 - (a_k/2^k)^2)}))}.$$

Write a program to display values for $k = 1, 2, 3 \ldots$ of a_k, using the alternative methods, and comment on the results.

5.6. Write a program to produce estimates of

$$\int_0^1 x^n e^{x-1} \, dx$$

using the backward and forward recursions described.

6 More on Entering Data and Displaying Results

6.1 Introduction

In all the examples we have considered so far, the information which has been transferred into or out of the program during its execution has been transmitted to or from the terminal. This is not always going to be convenient, particularly once we start handling large data sets or require a large number of results from a program.

In such instances we will want to be able to read into the program data that has already been stored in a file either via the computer's editor or as output from an earlier program. We will also want to send results and other information to a file so that we can preserve them for printing or perusal on the terminal at our leisure.

We have already met the concept of a file on the storage device linked to our computer as an entity in which the text or the compiled version of our program can be placed so that it can be preserved for future use. We now want to use such entities as storage for data and results from a program. Each file on the storage device has a filename which is unique. That is, only a single file with that name can exist. We will refer to this as the external filename.

Within a Pascal program we can also have a file of information in which data may be stored. This also has a filename that only exists when the program is running. We will refer to this as the internal filename.

Thus there are two uses of 'filename' which we must clarify. We also need to establish ways of linking between the two. There is

(1) the filename of the file that resides on the storage device of the computer that we are using, and
(2) the filename of the file of information stored within the memory of the computer which exists only when the program is running.

This latter facility is very useful in applications which involve handling large amounts of data temporarily while a program is being executed.

We will concentrate in this chapter on the problem of supplying data to a program and sending results to a file. The method for establishing the links between the two types of file is dependent on both the computer and the version of Pascal being used, and hence will not be discussed in detail here.

6.2 Output to Files

Information from a program is output using a WRITE or WRITELN statement. To direct it to a file we must inform the computer to which file we want to send the information. This is done by giving the internal filename as the first identifier in the WRITE or WRITELN statement. For example

WRITE(filename,list of information to be output);

This is consistent with our earlier use of these statements to send information to the screen since OUTPUT is the filename to which we are sending the information and the command

WRITE(list of information to be output);

is interpreted as

WRITE(OUTPUT,list information to be output);

As with other identifiers used in the program, the internal filename must be defined at the declaration stage of the program so that the compiler knows how to interpret it. Thus, if the first parameter of the WRITE or WRITELN is not a specified filename but a declared variable, or text contained in quotes, then the output is directed to the screen.

We will initially restrict ourselves to files of characters, called text files, since these can contain integer and real values as well as text. The filename is defined to be either of type TEXT or FILE OF CHAR (where CHAR is short for character). The filename must also be included as a parameter in the PROGRAM statement at the top of the program.

Before output can be stored in internal files, the file must be initialised so that it is ready to receive information. This is done using the REWRITE command. In standard Pascal this takes the form

REWRITE(internalfilename);

This would create an internal file which is ready to receive information but only exists while the program is being run. To save the file, we must assign an external name to it which is a process that depends on the computer and compiler that is being used.

The following program illustrates the definition of an internal file and writes a single line to it.

```
PROGRAM example9 (resultfile,OUTPUT);
VAR length : REAL;
    resultfile : TEXT;
BEGIN
    REWRITE(resultfile);
    length := 10.4;
    WRITELN(resultfile, 'length is ', length:6:2);
END.
```

Within the program, the output file is called *resultfile* which must be linked to an external file if the information is to be preserved. If we run the program it would create a file containing the single line

length is 10.4

Note that we have still included an OUTPUT in the PROGRAM statement, even though we do not actually direct any output in the program to the terminal. This is to ensure that any error messages that occur are displayed on the terminal. It should, however, be noted that some compilers do not insist on the inclusion of OUTPUT in the program statement.

Each WRITE or WRTIELN statement used in the program appends information to that already sent to the file referred to in the statement.

6.3 Input from Files

To read information from a file we use the READ or READLN statements, indicating the filename in the same way as the WRITE and WRITELN statements. The standard form is therefore

READ(filename,list of information to be read);

For example

READ(datafile,a,b,c);

would read values into a, b and c from an internal file called *datafile*.

Either READ(INPUT,a,b,c); or READ(a,b,c); where the filename is omitted, can be used if the information is to be read from the terminal.

An internal file must be initialised (RESET) so that information can be read from it starting from the beginning, using the standard Pascal command

RESET(internalfilename);

Data is read from a file sequentially. However, a file may be reset at any stage during a program so that reading can recommence from the beginning.

Since it is intended to link the internally declared file to one existing on the storage device, then the internal filename must also be placed in the list of parameters in the PROGRAM statement.

On the execution of the READ statement, no check can be made that these are the values you actually intended for a, b and c. That is the programmer's responsibility. The program will just read in the next three numerical values and assign them to the variables in the order specified by the list. However, certain general checks are made. For example, if an identifier in the read list was defined as an integer and the next piece of information in the input file was a real, then a runtime error would result. To make it clear where one number ends and the next begins a space must be left between the numbers, otherwise confusion will occur. For example, if we had a file from which we wanted to read an integer followed by a real and the file contained the two numbers 101 and 4.63 not separated by a space in the form 1014.63, then the program would read 1014 into the integer variable. The next character is the decimal point, and since real numbers must have at least one digit before the decimal point an error would result.

When entering data into a file using your computer's editor, the insertion of the necessary spaces is easy to achieve; but, if the information is created by an earlier Pascal program, then care must be taken to ensure that the spaces do in fact occur between the numbers.

A file may be of any length up to the maximum allowed by the computer. Each file contains an end-of-file marker which is the point beyond which there is no more data. When reading a file this point can be recognised by the standard BOOLEAN function EOF, which requires a single parameter, the filename.

EOF (filename)

It has the value TRUE if there is no more data in the file and FALSE otherwise.

We have seen how the use of WRITELN, when sending information to the terminal, causes the subsequent display to start a new line. When writing to a file it causes an end-of-line marker to be inserted in the file. This is used by the computer to signal a new line when we either list the file on the terminal or send it to a printer. When we try to read the information in the file into another Pascal program, it is treated as a separator between two numbers or as a space if read as a character. However, it is also detected by the special BOOLEAN function EOLN. When an end-of-line marker is reached while reading from a file, EOLN(filename) is set to TRUE, otherwise it has the value FALSE. It can be used in a fashion similar to that use of EOF in our earlier example.

Suppose we have, for example, a file containing values from several samples, each sample being stored on a separate line, then we could read the values in and calculate the means using program *example10* given below.

To enable us to move to the next line from the end of a line, we have to skip the end-of-line character which, apart from setting the BOOLEAN function EOLN, is treated as a blank. We can do this by the command

READLN(filename);

which skips to the beginning of the next line. Failure to include this command would mean that EOLN would remain TRUE and we would not move towards the next line. In a situation such as that in the next example the EOF marker would not be reached, leaving our program looping indefinitely.

The procedure READLN can also be used to read values into a list of identifiers in the same way as the READ statement. The difference is that when it has read values into the specified list it skips to the beginning of the next line, omitting any other values that remain in the file on the line that was being read.

The next **READ** or **READLN** command reads from the beginning of the next line in such a case.

```pascal
PROGRAM example10 (datafile,resultfile,OUTPUT);
VAR n,samplenumber : INTEGER; mean,next : REAL;
    datafile,resultfile : TEXT;
BEGIN
   RESET(datafile);
   REWRITE(resultfile);
   WRITELN(resultfile);
   samplenumber := 0;
   WHILE NOT EOF(datafile) DO
     BEGIN
       samplenumber := samplenumber + 1;
       mean := 0;
       n := 0;
       WHILE NOT EOLN(datafile) DO
         BEGIN
           READ(datafile,next);
           mean := mean + next;
           n := n + 1
         END;
       IF n = 0 THEN
         WRITELN('There is no data on line no ',
                                       samplenumber:1)
       ELSE
         BEGIN
           mean := mean / n;
           WRITE(resultfile, 'sample number',
                                     samplenumber:1);
           WRITE(resultfile, 'has ', n:1,
                                      'observations');
           WRITELN(resultfile, 'and mean ', mean:8:4)
         END;
       READLN(datafile)
     END
END.
```

We illustrate the use of EOF on its own in the next example where we calculate the mean of a set of values read from a file containing an unknown number of observations.

```pascal
PROGRAM example11 (datafile,OUTPUT);
VAR n : INTEGER; mean,next : REAL;
    datafile : TEXT;
BEGIN
   RESET(datafile);
   mean := 0;
   n := 0;
   WHILE NOT EOF(datafile) DO
     BEGIN
       READ(datafile,next);
       mean := mean + next;
       n := n + 1;
     END;
```

```
    IF n = 0 THEN
        WRITELN(' there is no data in the file ')
    ELSE
        BEGIN
            mean := mean / n;
            WRITE('the mean is ',mean:6:2);
        END;
END.
```

This may seem a straightforward application of EOF. However, there is unfortunately a possible problem with EOF because of the way it has been implemented on most versions of Pascal. The end-of-file marker can be thought of as an imaginary marker that comes at the beginning of a line. Thus, when the last observation has been read we are left at the end of the last line and not moved on to the next line, hence the end of file is not detected. Therefore we need some way of ensuring that we move on to the next line. One way is to check whether after each observation has been read we are at the end of a line, and if so we can move on to the next line using a READLN statement. Thus, after the read statement in the WHILE loop we insert the following statement:

IF EOLN(datafile) THEN READLN;

The use of files is important and becomes increasingly necessary as we develop our programming skills and tackle more complex mathematical problems. Bearing this in mind, future programs will be written so that the output is sent to a file for later display.

6.4 Construction of Tables

The more complex and extensive the output is from a program, the greater the need for careful planning of the way it is to be displayed on the screen or page — the main objective being to make the information easy to read.

Since in many situations we are asking the computer to produce the results from repeated operations we very quickly generate reams of results, not all of which we require and certainly more than we may wish to record by hand. It is, therefore, useful to consider the problem of controlling our output in order to produce neat, compact and easy-to-read tables of results.

6.4.1 Tables for comparison of results

For example, in section 4.3.1 we have a table which summarises the results from two methods of calculating $\exp(x)$. The table consists of a heading followed by a line of results for each value of x. This is achieved by the following steps:

(1) Display the heading.

   The spacing between column headings will usually need adjustment once the display in the main part of the table has been established, since the width of a column of results is determined by the numerical entries in it.

(2) For each *x* value

   calculate the various estimates and then display them in the order in which they are required.

Thus, in our example, a heading can be created by the statements

```
WRITELN('x':6,'EXP(x)':12,'method 1':16,
                         'method 2 (using n=12)':26);
WRITELN(' ':22,'approximation',
                         'n':3,'approximation':20);
```

while the numerical values can be displayed for each value of *x* in turn by including the following statement:

```
WRITELN(x:4:1,EXP(x):16:10,ans1:16:10,n:6,ans2:16:10);
```

Proceeding in this way a neat table is produced summarising the results without, temporarily, having to store the results in the computer's memory.

If the methods had been taken in turn, and the full set of estimates calculated, using one method, before moving on to the next method, we would either

(1) have had 60+ lines of output rather than 20, and the results would not have been so easy to compare, or

(2) have had to store the values in a set of identifiers, to have been displayed once all the calculations had been completed.

It should be noted that the output shown from using a :16:10 format for the REAL quantities does not result in the same amount of information being displayed about each number. If we had used an E format using :16, we would have used the same width of display for each number and have given each to 10 significant figures. Such a method would also have ensured that as e^x increased for larger *x* extra character positions were not allocated to the number, thereby destroying the alignment of the table.

6.4.2 *Tables of values*

Another common requirement is to produce a list of values. If the number of values to be displayed is small, a single line or column of values presents no problems. If, however, there are a large number, it is sensible to print them or display them in the form of a table.

For example, suppose we have a method to calculate

$$0.5 + \frac{1}{2\pi} \int_0^x \exp(-t^2/2)\, dt$$

for x=0.1 to 2.5 in steps of size 0.1. The values are then to be displayed as x value followed by the integral value in a neat table.

We will discuss methods for estimating such integrals in chapter 10.

The x values will be real values and need to be displayed with one decimal place. Hence :3:1 would be a format which is sufficient to display these values. The integral, which represents a probability from a normal distribution, takes values which lie between 0 and 1 and can be displayed to as many decimal places as required. A sensible display will be obtained using four decimal places.

The number of significant digits displayed for any number depends on the application but a general rule worth keeping in mind is that the human mind is rarely capable of distinguishing beyond the third or fourth significant digit. To display our result with four decimal places we need a format :6:4. If the non-decimal part of the number had contained more digits, the value of 6 would have to be increased accordingly. If the numbers are of variable magnitude, then the E format of display should be used to produce a more controlled output, using an equal number of positions on the screen or page for each number. To display four significant figures, for example, a format :10 would be necessary.

To separate the numbers a space is required. This can be achieved by a write statement of the form

```
WRITE(' ',x:3:1,' ',integral:6:4);
```

with the spaces included as text or by increasing the first value of the format information by one as follows:

```
WRITE(x:4:1,integral:7:4);
```

This last method adds a blank to the left of each number in order to fill the allocated positions.

The display of each set of information requires 11 character positions. Thus, if we had 80 characters on a line of the display medium, we could display information relating to up to seven x values, while if we had 132 positions we could display 12. In general, if we have n character positions and each piece of information requires m positions, then we can display up to (nDIVm) (that is, the nearest integer below n/m) pieces of information on each line.

The number displayed depends on a sensible consideration of the problem being solved. In this particular example, the result for five or ten x values per line has a nice symmetry and makes the table clearer to read than that for seven or twelve.

Suppose we want to display *k* pieces of information per line. After the *k* pieces of information have been displayed, we must issue a WRITELN statement which moves the output on to the next line by issuing a carriage return command. Thus, each *x* value and the corresponding integral value are determined and then displayed using a WRITE statement. A count is kept of the number of values displayed and once this has reached *k* a WRITELN statement is issued, moving us on to the next line, and the counter is reset to zero.

The structure of the program is as follows:

```
set the counter to zero
REPEAT
      Calculate the next value of x and the value of the integral
      Display the values using a WRITE statement
              For example WRITE(' ',x:3:1,' ',integral:6:4);
      Add one to the counter
      IF counter = k   THEN
              move the display onto the next line using
              a WRITELN statement
              set the counter to zero
UNTIL all the required values have been displayed.
```

The result of running the program is to produce output of the following form if *k*=5:

```
0.1 0.5398 0.2 0.5793 0.3 0.6179 0.4 0.6554 0.5 0.6915
0.6 0.7257 0.7 0.7580 0.8 0.7881 0.9 0.8159 1.0 0.8413
1.1 0.8643 1.2 0.8849 1.3 0.9032 1.4 0.9192 1.5 0.9332
1.6 0.9452 1.7 0.9554 1.8 0.9641 1.9 0.9713 2.0 0.9772
2.1 0.9821 2.2 0.9861 2.3 0.9893 2.4 0.9918 2.5 0.9938
```

6.5   Exercises

6.1. Write a program to produce a calendar for any year using Zeller's congruential method described in exercise **3.8** to determine the day on which January 1st falls.

6.2. Rework all the exercises in chapter 4 so that concise tables of results are produced and saved in a file.

7 Procedures and Functions

7.1 Introduction

In chapters 3 and 4 we introduced the idea of writing a program by breaking the problem down into a series of steps which we then convert into programming language. As problems become more complex, it is better to break them into a hierarchy of steps. In doing this we describe a problem using a structure of blocks, each of which is subdivided into smaller blocks until we have units small enough to handle in terms of the programming language.

It is therefore useful to be able to write our program in such a way that it reflects the structural approach to its development, and each block in our problem is a separate unit which can be programmed independently of the others.

In this chapter we develop this approach to programming by introducing the constructions in the Pascal language that allow us to write and test separately the parts of the program relating to particular blocks in our solution's structure.

It is also sometimes useful to be able to use a section of the program repeatedly and at different positions in the program. We can achieve both these aims by entering each block in the problem structure as a separate unit. In Pascal, these units are defined at the top of the program before they are used. They are given a name, and we use that name to access the defined unit at the required points in the program. There are two types of named unit in Pascal – *procedures* and *functions* – which we can use in this way. This useful way of structuring the program has the added advantage that when read at a later stage, it is easy to get an overall picture of the various tasks contained in the program. Constructing a program in this way follows naturally from the way a problem can be broken down into separate tasks.

7.2 Procedures

The declaration of a procedure is made after all the other declarations for the main part of the program but before any statements of the main part of the program. Structurally, a procedure is similar to the main part of the program and is in fact a miniprogram in its own right. There are two main features that differ:

1. Its heading begins with the reserved word PROCEDURE.
2. Its final END is followed by a semicolon rather than the full stop that
   indicates the end of the main program.

The name following the word PROCEDURE is the name by which the
procedure is called when the code it contains needs to be used.

The execution of the program follows the steps in the main BEGIN. . .END
block which occurs at the bottom of the program, using the code supplied in
the procedure definition when required.

Example

Consider the following program which illustrates the use of a simple procedure
to calculate the area of a circle.

```
PROGRAM example12 (INPUT,OUTPUT);
VAR area,radius : REAL;
CONST pi = 3.1416;
PROCEDURE circlearea;
   BEGIN
     area := pi * radius * radius;
   END;
BEGIN
   WRITELN('Type in the radius');
   READ(radius);
   circlearea;
   WRITELN('The area of a  circle  with radius',
                          radius:6:2, ' is',area:6:2);
END.
```

The procedure is called *circlearea* and is declared before it is used in the program.
Note that the variables *area* and *radius* declared in the main part of the program
are available for use in the procedure. All quantities declared in the main part of
the program are called *global* quantities.

It is also possible to declare variables and constants within the procedure for
use solely within the procedure. Such quantities defined within a procedure are
called *local* quantities and can only be used within the procedure. The values
held by such variables or constants are lost after execution of the procedure,
and are not available for further use when the next step in the main program is
being obeyed. Their values, as previously set, are therefore not available when
the procedure is re-entered.

Global variables, declared in the main part of the program, have values which
may be used in procedures, and any values they take during the execution of
the procedure are available for use as the main program progresses (that is, after
completion of execution of the procedure).

If there is a clash of names — that is, an identical variable name is used in
both the main program and as a local variable in a procedure — then when the

procedure is being executed the variable behaves as a local variable and does not alter the value of the global variable in the main part of the program.

Note: any variable that is used as a control variable in a **FOR** loop in a procedure must be defined as a local variable.

Example

The following example illustrates the use of a supplied procedure to calculate the cosine of a supplied argument using the MacLaurin expansion:

$$\cos(x) = 1 - \frac{x^2}{2!} + \frac{x^4}{4!} - \frac{x^6}{6!} + \ldots$$

It should be noted that a cosine function, $COS(x)$, is already available in Pascal but the example is used here for illustrative purposes. For this reason we use the name *cosine* rather than *cos* which is the name of the function.

```
PROGRAM cosinecalculation (INPUT,OUTPUT);
VAR sum,x : REAL;
PROCEDURE cosine;
   CONST epsilon = 0.0001;
   VAR term : REAL; i : INTEGER;
   BEGIN
     i := 2;
     sum := 1;
     term := 1;
     REPEAT
       term := - term * x * x / (i * (i - 1));
       sum := sum + term;
       i := i + 2;
     UNTIL ABS(term) < epsilon;
   END;
BEGIN
   READ(x);
   cosine;
   WRITELN('the cosine of ', x:8:4, ' is ', sum:8:4);
END.
```

The variables *sum* and *x* are declared in the main program and their values are available and can be used or altered in both the main program and the procedure. The constant *epsilon* and the variables *term* and *i* are declared in the procedure and hence can only be used within the procedure.

7.2.1 Parameters

The program *cosinecalculation* consisted of a procedure that is called simply by its name, and any values generated within it are passed to the main part of the program by the use of global variables. This allows us to use the *same* operation (that is, calculate $\cos(x)$) at different parts of the main program. However, it is also likely that we would like to perform *similar* operations at different parts of the program.

We might, for example, want to use the cosine procedure to calculate the cosine of values stored in identifiers y and z. Using the routine as it stands, we would have to assign the value of each variable to the global identifier x and then, having used the procedure to calculate the cosine, assign the result, which is in the global identifier *sum*, to another identifier.

Fortunately, Pascal allows us a neater way of doing this using parameters in the procedure call. In the above example there are two ways we can use the parameters. Firstly, we want to be able to pass the value of the identifiers y and z into the procedure so that they are assigned to the identifier x and, secondly, we want to extract the result from *sum* and assign it to identifiers *cosineofy* and *cosineofz*, say, respectively. The first use is an example of a *value parameter* which we can use simply to pass a value into the procedure, while the second is an example of a *variable parameter* which extracts the information from the procedure.

We could therefore redefine our procedure by including the parameters in the heading of the procedure as follows:

```
PROCEDURE cosine(x : REAL; VAR sum : REAL);
```

This defines two variables x and *sum*, the first of which can only accept values while the second can be used to transfer information, in this case the cosine of x, from the procedure to the main program. When this procedure is called we can use identifiers, defined in the main program as the parameters. For example, we could use cosine(y,cosineofy) provided y and *cosineofy* have the same type as the parameters. A value parameter, since it assigns a value to the identifier, can be supplied information by an expression that is acceptable as the right-hand side of an assignment statement. Thus, for example, cosine(2*pi/n,cosineofy) or cosine(4.324,cosineofy) are acceptable. The variable parameter, though, must be an identifier since it assigns a value to this identifier. Thus, only terms acceptable on the left-hand side of an assignment statement are acceptable. Thus, for example, cosine(x,3.4) or cosine(3.4,x/y) are unacceptable calls.

When a procedure is called, the parameters must be listed in the same order as they have been defined. Also, any variables defined in the PROCEDURE statement cannot be declared at the declaration stage of the procedure definition.

The following example illustrates the use of parameters using a rewritten cosine procedure.

```
PROGRAM cosinecalculation2(INPUT,OUTPUT);
VAR y,z,cosineofy,cosineofz,ans : REAL;
PROCEDURE cosine(x : REAL; VAR sum : REAL);
   CONST epsilon = 0.0001;
   VAR term : REAL; i : integer;
   BEGIN
      i := 2;
      sum := 1;
      term := 1;
      REPEAT
         term := - term * x * x / (i * (i - 1));
         sum := sum + term;
         i := i + 2;
      UNTIL abs(term) < epsilon;
   END;
BEGIN
   WRITELN('What are the values of y and z ?');
   READ(y,z);
   cosine(y,cosineofy);
   cosine(z,cosineofz);
   ans := cosineofy + cosineofz;
   WRITELN(' the answer is ',ans:6:2);
END.
```

Note that a variable parameter could also be used to supply information to the routine as well as extract it. In other words, the value that the parameter has on entry can be used in the procedure before it is changed within the procedure.

7.3 Functions

If a procedure is used for calculating a single value, then an alternative method is to define it as a function. We have already met and used standard functions such as SQRT, EXP, LN etc. and any function we define ourselves is used in the same way. A function is defined in a similar way to a procedure and obeys the same rules regarding the use of global and local variables. The section 7.2.1, on parameters, is again relevant. There are, however, two important differences between functions and procedures:

(1) The function must have a type. This is specified in the function declaration after the list of any parameters, the heading of the function declaration having the following form:

FUNCTION name (specification of any parameters):type;

For example

FUNCTION max(a,b:REAL):REAL;

heads a function, called max, which is of type real and has two real parameters. The result of using the function is to produce a real result.

(2) The function must also have a statement somewhere within it, usually at the end, which assigns a value to the name. Thus, the above example must have a statement beginning max:= within its set of statements.

Care should be taken by the beginner not to use the name of the function on the right-hand side of an assignment statement in the function declaration. In the context of Pascal this implies a recursion in the sense that the function calls itself when its name is encountered on the right-hand side of an assignment statement. In the hands of the inexperienced programmer this is likely to cause problems such as the program repeating an operation indefinitely. However, this may not show as an error when the program is compiled or run since it is a valid construction and can be very useful in more advanced applications.

The following example illustrates the use of a function to determine the maximum of two values.

```
PROGRAM exampleofusingafunction (INPUT,OUTPUT);
VAR x,y,z,ans : REAL;
FUNCTION max(a,b : REAL) : REAL;
  BEGIN
    IF a < b THEN max := b
    ELSE max := a;
  END;
BEGIN
  WRITELN('What are the values for x,y and z ?');
  READ(x,y,z);
  ans := max(x,y);
  ans := max(ans,z);
  WRITELN('the maximum is ',ans:6:2);
END.
```

As we have done with pre-defined functions, introduced in chapter 2, we can use the function as an argument so we could have replaced two lines in our main program by the line

```
ans := max(max(x,y),z);
```

and achieve the same result.

Our cosine PROCEDURE could also be written as a FUNCTION since it returns a single value to the calling program. This is illustrated in the following function in which the alterations from the original procedure are indicated in italics.

```
FUNCTION cosine(x:REAL) : REAL;
  CONST epsilon = 0.0001;
  VAR term : REAL;i : integer;
      sum : REAL;
```

```
BEGIN
    i := 2;
    sum := 1;
    term := 1;
    REPEAT
      term := - term * x * x / (i * (i - 1));
      sum := sum + term;
      i := i + 2;
    UNTIL abs(term) < epsilon;
    cosine := sum;
END;
```

We note that now only a single parameter is required. We should also note, however, that the real variable *sum* is still required in the form of a local variable. This variable is used in the accumulation of the series and it is only once the calculation is complete that we assign the value to the function name. If we had used the identifier *cosine* earlier, in place of *sum*, then the statement in which we add the series would have had the term *cosine* on the right-hand side of the assignment. In this situation it would have called the function itself and so on. Clearly, in this case an error would have resulted since the cosine call on the right-hand side of the assignment does not have the necessary parameter for the function.

7.4 Nesting of Procedures and Functions

It is also possible to define procedures and/or functions within procedures or functions, building up a hierarchy of block structures. Consider the following program structure.

```
PROGRAM nestingexample(INPUT,OUTPUT);
VAR a,b : INTEGER;
PROCEDURE outer;
  VAR c,d : INTEGER;
  PROCEDURE inner;
    VAR e,f : INTEGER;
    BEGIN
      (* statements for procedure inner *)
    END;(* end of inner *)
  BEGIN
    (* statements for procedure outer *)
  END;(* end of outer *)
BEGIN
    (* statements in the main part of the program *)
END.(* end of the main part of the program *)
```

In this example of program structure, variables *a* and *b* are global to the whole program and can be used in both procedures as well as the main program.

Variables c and d, defined in procedure *outer*, are local to that procedure. They can, however, be used in procedure *inner* since they are considered global within procedure *outer*. They cannot be used in the main part of the program.

Variables e and f, defined in procedure *inner*, are local to that procedure and cannot be used elsewhere.

The procedure *inner* cannot call the procedure *outer* since it is local to *outer*. Procedure *outer*, on the other hand, can call the procedure *inner*.

The main program can call the procedure *outer* since *outer* is local to it. On the other hand, it cannot call the procedure *inner* since *inner* is a local procedure of *outer*.

We can summarise the rules that procedures and the variables defined within them must obey as follows:

1. A procedure behaves as a program in its own right.
2. Variables defined in a procedure are available throughout that procedure.
3. Procedures can call procedures defined before them in the same procedure. That is, if two procedures are defined as local to a third procedure then the one that is defined second can call the first but not vice versa.
4. Procedures defined within procedures are local to that procedure and cannot be called directly from outside that procedure.

7.5 Structured Program Design

We have seen how, for simple problems, we can design the program by breaking the problem into steps rather than attempting to translate it straight into code. For more complex problems, it is sensible to introduce a hierarchy of component parts so that the problem is divided into blocks, each of which can also be sub-divided into blocks and so on. For example, the majority of computational problems can be broken into three stages at the first level, namely

1. Input of information
2. Calculation
3. Output of results.

Each of these blocks could be written as procedures so that the main block of a program could in its simplest form consist simply of calls to these procedures as follows:

```
BEGIN
    inputdata;
    calculate;
    outputresults;
END.
```

The steps are immediately clear to the reader. The various stages could also be broken down into other blocks so that they are easy to follow. The other advantage of building a program in this way is that the procedures can be tested separately before they are included in the main program, thereby giving us more confidence in the final program.

The example above may be extreme, but structuring programs in this way does make them easier to read, write and correct.

Breaking a problem down stage by stage also helps us to develop a good programming style which results from a clear understanding of any problem and a correct translation from problem to program. It also results in the final program reflecting the structure of our solution to the problem.

The majority of programs in Pascal can be written using about half a dozen programming structures. These can be combined by using the three concepts of repetition, decision and modularity.

We have already met various programming structures as follows:

1. The single statement, which is the simplest.
2. The various control statements.
   FOR, WHILE and REPEAT loops which allow us to repeat blocks of statements.
   IF structures which allow us to select from various sets of statements.
3. Procedures and functions.
   These allow us to fully structure a program by separating the code for various operations from the main part of the program.

The form of these constructions encourages us to break a problem down into blocks of operations and helps in the development of carefully thought-out programs.

7.5.1 Modular construction

To encourage the use of structuring in our programs, we can introduce a hierarchy of blocks in the representation of our problem. Each block represents a part of our program and can be subdivided to indicate the steps used within that stage. Using this procedure it is sensible to program the various blocks as procedures and functions, thus separating the writing of these parts of the code from the main part of the program.

Once we have an idea of the basic structure of the algorithm which we are going to use for our program and have broken down each block into its steps, only then should we begin to consider the identifiers and data structures that we are going to use in our program. Following this process and building up the program slowly can help to produce a good and accurate program. It can also avoid many errors that can be introduced by attempting to write the program while sitting at the computer.

We can illustrate the use of modular structuring by considering how we can construct a program to produce a table of numerical results for comparing different methods of calculating $\exp(x)$. In chapter 4 we described a method of calculating exponential functions using the MacLaurin expansion and in chapter 5 we discussed how table 4.1, in which the results from the two alternatives are summarised, could be produced.

Let us now consider how to construct this program.

The basic building blocks are:

(1) Display the heading of the table.
(2) Calculate and display the numerical values that make up the rest of the table.

The first block is fairly straightforward and can be carried out by using a simple sequence of WRITE and/or WRITELN statements.

The second block consists of a repeated set of operations as follows:

    (2)   For each x value
        (a) calculate e^x using the various methods
        (b) display the numerical results.

2a consists of two separate blocks, namely

    (i) calculate e^x using a fixed number of terms
    (ii) calculate e^x using a stopping condition

Each of these can now be written as separate procedures called at the appropriate place. The structure is summarised in figure 7.1.

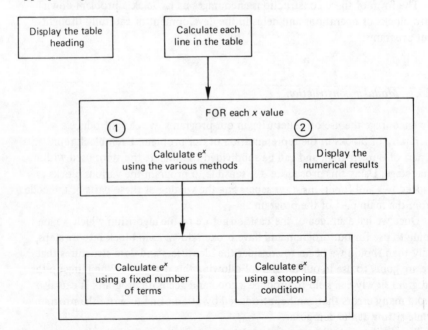

Figure 7.1 Structure of program for calculating $\exp(x)$ by two methods

If we now wished to include the improved version for comparison, we simply insert a third calculation block at step 2a. We have already seen in chapter 4 that the improved version itself consists of a three stage process:

reduce the exponent   calculate exp(x)   power up to regain the ex-
to the range   using the Maclaurin ponential of the supplied
$-0.5, 0.5$ expansion   exponent

The advantage of this block structure approach is that we can construct the various parts of the program separately and test them in small programs before using them in larger and more complex ones. We can also build up a library of procedures and techniques that can be used almost off the shelf, thus making it easier to construct programs in the future.

The final block structure is shown in figure 7.2.

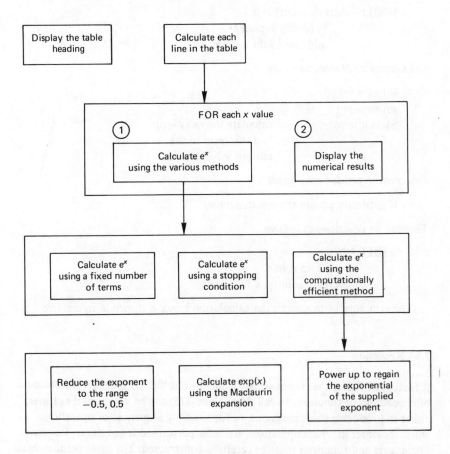

Figure 7.2   Revised structure showing how the efficient method is added to the program

If we now take the computationally efficient method, we can illustrate how this is broken down stage by stage until we have the final structure where each block represents a statement in the program. Thus, each of the three steps can be broken down as follows:

Reduction of exponent

> set counter k to zero
> if necessary reduce the exponent until
> > it is in the range -0.5 to 0.5
> > and record the number of steps
> > in the counter k

This second step can now be broken down still further as follows

> WHILE abs(exponent) > 0.5
> > halve the exponent
> > add one to the counter k

Calculation of Maclaurin series

> set sum = 1
> set term = 1
> SUM n terms   calculate the next term
>   add this term to the
>   current value of sum

Power up to produce the result

> if necessary square the results k times

This can be described as follows

> WHILE $k > 0$
> > square up sum
> > decrease counter k by 1

The structure of this particular calculation block is shown in figure 7.3.

7.6 Recursion

A feature of Pascal that is not widely available in other languages is the ability of procedures or functions to call themselves. This can be very useful at times, either in producing short programs, or for handling sequences of operations when the order in which operations are to be performed is not linear. Recursive procedures and functions must be carefully constructed. The main point to bear in mind is that the process must stop at some stage, otherwise the procedure or function will call itself *ad infinitum*. This can be simply achieved by ensuring

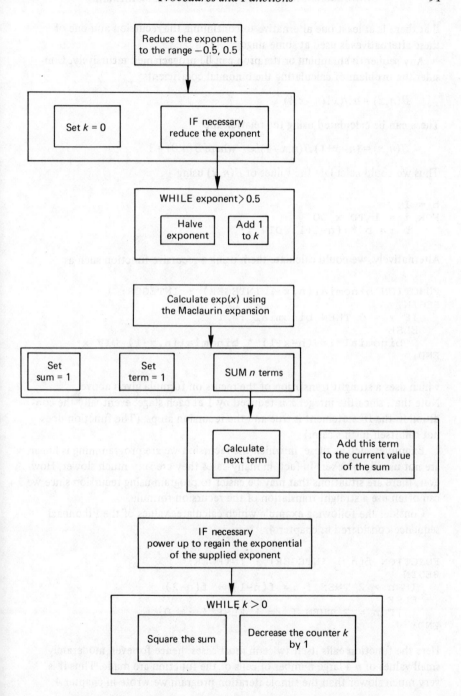

Figure 7.3 Structure for programming the efficient method

that there is at least one alternative to continuing the recursion and one of these alternatives is used at some stage.

Any sequential mapping or iteration can be programmed recursively. Consider the problem of calculating the binomial coefficients:

$$B(n,x) = n!/(x!(n-x)!)$$

These can be calculated using the relationship:

$$B(n,x) = (n-x+1) B(n,x-1)/x \quad \text{where } B(n,0) = 1$$

Thus we could calculate the values of $B(n,x)$ using

```
b = 1;
FOR i:= 1 TO x DO
    b := b * (n-i+1) DIV i;
```

Alternatively, we could calculate them using a recursive function such as

```
FUNCTION binomial(n,x : INTEGER) : INTEGER;
BEGIN
    IF x = 0 THEN binomial := 1
    ELSE
       binomial := (n-x+1) * binomial(n,x-1) DIV x;
END;
```

which uses a straight translation of the recursion formula given above.
Note that, since the integer x is reduced by 1 at each stage, eventually the condition in the IF statement is true and the recursion stops. (The function does not call itself in this case.)

Examples such as these, in which the recursion we are programming is linear, are not usually quicker. In fact, in many cases they are very much slower. However, there are situations that may be easier to program using recursion since we can often use a straight translation of the recursion formula.

Consider the following example which calculates values of the Fibonnaci sequence considered in chapter 4.

```
FUNCTION f(n : INTEGER) : INTEGER;
BEGIN
    IF n > 2 THEN f := f(n-1) + f(n-2)
    ELSE
       IF n = 2 THEN f := 1 ELSE f := 0;
END;
```

Here the function calls itself twice in most cases, hence for even moderately small values of n a large number of calls of the function are made. Thus it is very much slower than the simple iteration program we wrote in chapter 4.

However, in situations where the recursions are not carried out in a linear order, as well as being a neater way of writing the program it can also be much quicker. We will meet some examples of these in chapter 20 when we consider trees and graphs.

7.7 Exercises

7.1. Write a program that contains three functions for evaluating SINH, COSH and TANH respectively, and uses these functions to tabulate SINH(x), COSH(x) and TANH(x) for values of x from 0 to 3.4 in steps of 0.2.

7.2. Given any two positive integers, there exists a largest integer d which divides both these integers. This value d is called the greatest common divisor (gcd) or highest common factor (HCF).

Given two positive integers m and n then, the Euclidean algorithm for calculating d is given by the steps:

(a) Calculate the remainder m/n where $m>n$.
(b) Replace m by n.
(c) Replace n by the remainder.
(d) If the remainder is not equal to zero then go back to step 1.
     If the remainder is zero then the current value of m is the gcd
     (that is, the last non-zero remainder).

Write a procedure to calculate the gcd of positive integers and use this procedure in programs to (1) determine the gcd of three numbers and (2) determine if two positive integers are relatively prime (that is, their gcd is 1).

7.3. Write a function which raises an integer to an integer power.

7.4. Write a function to raise a real number to a real power.

7.5. Write a procedure to reduce a fraction represented as two integers to its lowest terms.

7.6. Write procedures to add and multiply two fractions. How would you use the procedures to handle the subtraction and the division of fractions?

7.7. Write a recursive procedure for determining the gcd of two numbers.

7.8. Run the programs described in chapters 4 and 7 to determine the values in the Fibonnaci sequence, comparing the length of time they take to produce n terms.

7.9. Write recursive functions to calculate the values of probabilities from the Binomial and Poisson distributions using the formulae given in exercises **4.7** and **4.8** respectively.

8 Iterative Procedures

8.1 Iteration

A problem which arises in many areas of numerical mathematics is one of find-ing the roots of an equation of the form $f(x) = 0$. The function is often non-linear and we usually cannot obtain an algebraic formula which would allow us to calculate the solution. We therefore need to use numerical methods, and in this chapter we consider some of the more obvious approaches to the problem and ways of programming them.

In chapter 4 we introduced the idea of a sequential mapping which can be used to produce a sequence of values starting from an initial value. We also showed that by rewriting the equation $x^2 + 2x - 1 = 0$ in the form $x=1/(2+x)$, then the sequence generated by the sequential mapping

$$x_{n+1} = 1/(2+x_n)$$

tends towards the positive root of the equation for nearly all starting values.

Any calculation of the form

$$x_{n+1} = g(x_n) \quad n = 0, 1, 2 \ldots$$

where we start from an assumed value x_0 and calculate $x_1, x_2 \ldots$ in succession, is called an iterative process. If, as n tends to infinity, x_n tends to a definite value then the iterative procedure is said to *converge*. If, on the other hand, x_n does not converge (for example, it tends to infinity) then the procedure is said to *diverge*.

If we had a linear equation of the form $3x-2 = 0$ then the solution is obtained by rearranging the equation as $x = 2/3$. We therefore begin by con-sidering the problem of non-linear functions by rewriting our equation in the form

$$x = g(x)$$

as we did in our earlier example. Using this as a sequential mapping $x_{n+1} = g(x_n)$, we have an iterative procedure which converges if $|g'(x)| < 1$ in the neighbour-hood of the root and the starting value is in this neighbourhood. We can illustrate this graphically as shown in figure 8.1.

If we had rewritten our original example as

$$x = (1-x^2)/2$$

then convergence can be guaranteed, to the positive root, for starting values in
the range $(-1, 1)$.

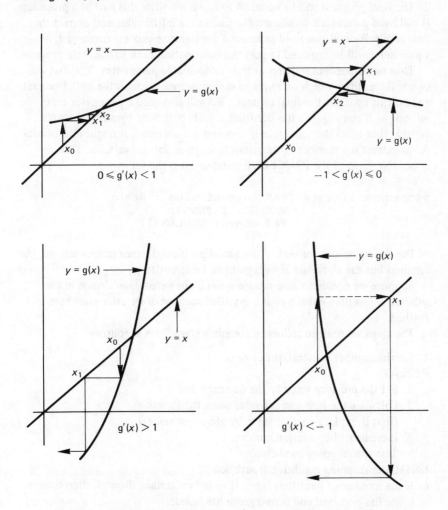

Figure 8.1 Performance of the iterative procedure $x_{n+1} = g(x_n)$

8.2 Programming an Iterative Procedure

As usual we begin by writing down the three basic steps:

1. Request the initial value.
2. Perform the iteration until the stopping condition is satisfied.
3. Output the result.

Steps 1 and 3 are reasonably straightforward and can be handled by techniques we have already covered. We concentrate, therefore, on developing step 2. The final program will be easier to follow if we write this part as a procedure. It will need parameters to allow us to pass in the initial value and extract the final result. We will also need to know if the iterative process converged, hence a parameter will be required to pass this information back to the main program.

Thus our procedure requires at least two variable parameters. The first will be a real identifier which on entry to the procedure contains the initial estimate and on exit contains the final estimate. We will also need a parameter that informs us if convergence has occurred. This is best done by a BOOLEAN variable that takes the value true if convergence occurred. It might also be useful to extract the number of iterations through an integer variable. Thus a reasonable form of the PROCEDURE statement is the following:

```
PROCEDURE iterate (VAR currentvalue : REAL;
                   VAR n : INTEGER;
                   VAR conv : BOOLEAN);
```

Depending on the context of the actual problem, further parameters may be required but the above are always going to be essential.

Suppose we decide to stop our iteration if the percentage change in the solution is less than epsilon or if a specified number of iterations has been reached.

The steps we need to follow to complete stage 2 are as follows:

1. Set the number of iterations to zero.
REPEAT
   2. Set the previous value to the current value.
   3. Calculate the new current value using the function
      (that is, replace current value by g(current value)).
   4. Increment the iteration number.
   5. Test the stopping condition.
UNTIL the stopping condition is satisfied
6. If the number of iterations is less than the maximum allowed, then convergence has occurred else convergence has failed.
   Set the BOOLEAN variable conv accordingly.

Before coding the procedure, we have to decide the best way to represent the calculation in step 3. This involves the use of the function $g(x)$. It can either be represented as the right-hand side of an assignment statement or as a FUNCTION declared above the procedure in the program. The latter is more useful since the same basic procedure can be used for other iterative processes. Thus we are going to use a declared function in step 3 of the procedure. In this case it is going to require a single parameter, namely the current value of the sequence, which is going to be a real quantity.

Translating this into Pascal we have a procedure of the following form.

```
PROCEDURE iterate (VAR currentvalue : REAL;
                   VAR n : INTEGER;
                   VAR conv : BOOLEAN);

(* Procedure to perform the iterations *)

CONST epsilon = 0.0001; maxn = 100;
VAR previousvalue : REAL;

BEGIN
   (*Step 1*)
   n := 0;
   REPEAT
     (* Step 2 *)
     previousvalue := currentvalue;
     (* Step 3 *)
     currentvalue := g(currentvalue);
     (* Step 4 *)
     n := n + 1;
     (* Step 5 *)
   UNTIL (ABS(previousvalue/currentvalue-1) < epsilon)
                 OR     (n = maxn);
     (* Step 6 *)
     conv := NOT (n = maxn);
   END;
```

The problem described in section 8.1 can now be solved using the procedure as follows:

Step 1. Read in the initial value.
Step 2. Perform the iterations.
Step 3. Display the results.

The function used in the iterative procedure must be placed above the procedure which calls it.

If we now translate this into Pascal, we obtain the following. The code for the procedure to perform the iterations is placed in the position indicated.

```
PROGRAM example13 (INPUT,OUTPUT);
VAR x : REAL; iterations : INTEGER;
    convergence : BOOLEAN;

FUNCTION g(x : REAL) : REAL;
  BEGIN
    g := 1/(2+x);
  END;


PROCEDURE iterate (VAR currentvalue : REAL;
                   VAR n : INTEGER;
                   VAR conv : BOOLEAN);
(* Insert the text of  the  procedure  to perform  the
iterations *)
```

```
BEGIN
   WRITELN(' what is your initial value ');
   READ(x);

   iterate(x,iterations,convergence);

   IF convergence THEN
        BEGIN
           WRITELN(' Convergence successful after ',
                           iterations:1,' iterations');
           WRITELN(' The final solution is ',x:10:4);
        END
   ELSE
        BEGIN
           WRITELN('Convergence failed after ',
                           iterations:1,' iterations');
           WRITELN('The final value was ',x:10:4);
        END;

END.
```

8.3 Newton–Raphson

We have considered the problem of determining a root of the equation

$$f(x) = 0$$

by rewriting the equation. We have illustrated that finding a solution depends on being able to rewrite the equation so that the absolute value of the derivative of the right-hand side is less than 1 in the neighbourhood of the solution.

Suppose we consider an alternative method based on using the derivative of the non-linear function f', thereby avoiding the need to rewrite the equation. We illustrate the method graphically starting from an initial guess x_0.

The derivative at x_0, $f'(x_0)$, is the slope of the tangent to the curve at x_0 and reaches the x-axis at the point x_1.

From figure 8.2 we see that

$$f'(x_0) = - f(x_0)/(x_1 - x_0)$$

hence $x_1 = x_0 - f(x_0)/f'(x_0)$

If we use this process to produce a sequence of points $x_0, x_1, x_2 \ldots$ where

$$x_{n+1} = x_n - f(x_n)/f'(x_n)$$

then we have an iterative procedure which also produces a solution. This method is faster, when it works, although poor initial values can cause problems.

Problems, for example, can arise when the value of $f'(x)$ at any particular value of x is close to zero. The result of this can be that the next estimate in the sequence is very poor.

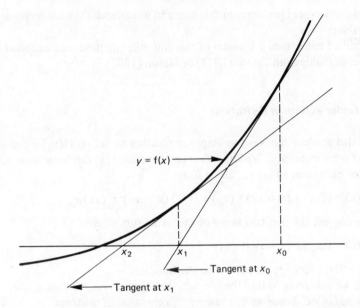

Figure 8.2

If there is more than one root to the equation then the solution is not neces-sarily the closest to our starting value, as is shown in figure 8.3.

If the problem has more than one root with the same value then the deriva-tive at the root is zero since it must be a turning point or point of inflection.

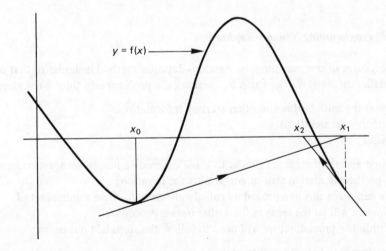

Figure 8.3

Hence we can expect problems in this case and in particular the convergence becomes very slow.

A detailed theoretical discussion of this and other methods can be found, for example, in Phillips and Taylor (1973) or Maron (1987).

8.3.1 Taylor expansion derivation

We can also produce the Newton–Raphson solution to our problem by considering the Taylor expansion. We know that any function $f(x)$ can be written, using its Taylor expansion about x_0, in the form

$$f(x) = f(x_0) + (x - x_0)\, f'(x_0) + (1/2)\, (x - x_0)^2 f''(x_0) + \ldots$$

If we use just the first two terms of this expansion then

$$f(x) = f(x_0) + (x - x_0)f'(x_0)$$

gives us a first-order approximation to the function.

Since we are trying to find the roots of the equation $f(x)=0$ then at the solution x, $f(x)=0$, hence we can rewrite this equation in the form

$$x = x_0 - f(x_0)/f'(x_0)$$

This equation is used to calculate the next term in a sequence. Since it is formed by an approximation to the original function then the first step will not lead us directly to the solution. We therefore need to use it as an iterative procedure and it generates the same sequence as suggested by the graphical approach.

8.3.2 Programming Newton–Raphson

The problem of programming the Newton–Raphson method is similar to that of the earlier method. We can begin by breaking the problem into three basic steps.

1. Enter the initial value and other start-up information.
2. Perform the iterations.
3. Display the results.

The first and third steps are similar to those covered earlier, so we again concentrate on the calculation step of performing the iterations.

We can place this in a procedure called *newtonraph* whose minimal set of parameters will be the same as the earlier *iterate* procedure.

Within the procedure, we will need to follow the steps laid out below:

1. Set number of steps equal to zero.
REPEAT
    2. Set previous value equal to current value.

3. Current value:=current value − f(currentvalue)/f′(current value).
4. Add 1 to the number of iterative steps.

UNTIL the stopping condition is satisfied.

5. Set convergence indicator to the appropriate value.

Now we have structured the problem, we see that we can write virtually the same procedure as before. The only difference is that the step which actually calculates the next term in the sequence requires two functions: one to supply the current value of the function and the other to supply the current value of the derivative of the function. These will both have at least one parameter for receiving the current estimate and both will be REAL functions. The form of the two functions depends on the actual problem that is being solved at the time.

Translating this into Pascal, we have the following procedure which is very similar to that for the earlier method.

```
PROCEDURE newtonraph (VAR currentvalue : REAL;
                      VAR n : INTEGER;
                      VAR conv : BOOLEAN);

(*procedure to perform the newton-raphson iteration *)

CONST epsilon = 0.0001; maxn = 20;
VAR previousvalue : REAL;

BEGIN

   n := 0;
   REPEAT
      previousvalue := currentvalue;
      currentvalue := currentvalue -
               f(currentvalue)/fdash(currentvalue);
      n := n+1;
   UNTIL (ABS(previousvalue/currentvalue-1) < epsilon)
                   OR  (n = maxn);

   conv := NOT (n = maxn);
END;
```

8.4 The Secant Method

The Newton–Raphson method is based on knowledge of the gradient of the curve at the estimate of the solution which is used to construct the tangent at that point. The next estimate is chosen to be the point where this tangent cuts the axis. The derivative may not always be easy to obtain, hence possible alternative methods should be considered. The curve between two points can be approximated by a straight line and without knowledge of the derivative we need two points on the curve to estimate this line.

Suppose we have initial estimates x_0 and x_1 at which we can calculate $f(x_0)$ and $f(x_1)$. Then the slope of the line, or secant to the curve, between these two points is given by

$$g = \frac{f(x_1) - f(x_0)}{x_1 - x_0}$$

Hence the equation of the line through $(x_1, f(x_1))$ is given by

$$\frac{y - f(x_1)}{x - x_1} = g$$

that is

$$y = f(x_1) + g(x - x_1)$$

The point where this line cuts the x-axis can be considered to be a revised estimate of the root of the equation $f(x)=0$; that is

$$x_2 = x_1 - f(x_1)/g$$

$$= x_1 - f(x_1) \frac{(x_1 - x_0)}{f(x_1) - f(x_0)}$$

This is demonstrated in figure 8.4.

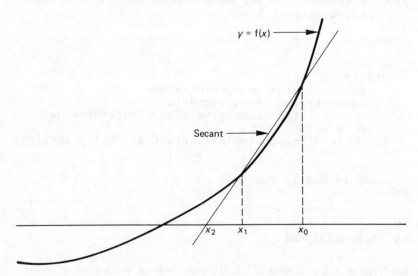

Figure 8.4

Replacing x_0 by x_1, and x_1 by x_2, and repeating the process we obtain a further revision of the estimate. Thus we have another example of a sequential mapping in the form

$$x_{n+1} = x_n - f(x_n)\ \frac{(x_n - x_{n-1})}{f(x_n) - f(x_{n-1})}$$

which is very similar to Newton-Raphson.

However, once the function has been evaluated at x_0 and x_1 we only need to evaluate the function once at each step of the iteration to obtain $f(x_2)$, $f(x_3)$ ... etc. Thus while it is possible that more iterative steps may be required compared with Newton-Raphson, only half the function evaluations are required which may mean that less time is required to complete the iterative process. This, together with the fact that the knowledge of the derivative of the function is not required, makes it worth considering as an alternative method.

8.4.1 Programming aspects

The basic structure of a procedure for the secant method is the same as for the other iterative methods considered since the same information is required on completion of the iteration steps. We will, however, require one extra parameter since two initial values are required. This extra parameter can be a value parameter since we only need to pass the final solution back to the main program. Suppose we call the parameters containing the initial values x2 and x1 with the final solution being passed back through x2.

Thus we have a procedure:

```
PROCEDURE secantmethod (VAR x2 : REAL; x1 : REAL;
                 VAR n : INTEGER; VAR conv : BOOLEAN);
CONST maxn = 100; epsilon = 0.0001;
VAR x0, grad0, grad1 : REAL;

BEGIN
  n := 0;
  grad1 := f(x1);

  REPEAT
(* store the latest two values in x0 and x1 *)
    x0 := x1;
    x1 := x2;
(* store the gradients of the current points *)
    grad0 := grad1;
    grad1 := f(x2);
(* extrapolate to the next point *)
    x2 := x1 - grad1 * (x1 - x0) / (grad1 - grad0);
    n := n + 1;
  UNTIL (ABS(x2 / x1 - 1) < epsilon) OR (n = maxn);

  conv := NOT (n = maxn);

END;
```

This can be used in the same way as the procedures given for the other methods.

8.5   Functions and Procedures as Parameters

There are likely to be many instances when we will want to use procedures, such as the above examples of iterative procedures, on two or more functions within a single program. As we have written our procedures using a function called f, if at a later time we want to use the procedure on a function g, say, we would have to supply a second procedure virtually identical to the existing one except for the name of the function that is called. This defeats one of the ideas behind using procedures and functions in the first place.

However, in Pascal, we can avoid this problem by declaring the function as a value parameter of the procedure and then, provided the functions have the same number of parameters as specified in the procedure definition, we can call the procedure first with f as a parameter and then with g. Thus we could modify the heading of our procedure *newtonraph* to read

```
PROCEDURE newtonraph(VAR currentvalue : REAL;
                     VAR n : INTEGER;
                     VAR convergence : Boolean;
                     FUNCTION f(x : REAL) : REAL);
```

The declaration of the function f as a parameter of the procedure has the same form as the heading of the functions that are to be passed. The name, type and its parameters must all be specified. The names of the parameters of the function, however, are arbitrary and only serve to hold a position for the parameters when the function is called.

When the procedure *newtonraph* is called at any stage in the program the particular function name, without parameters, is placed in the appropriate place. Thus

> newtonraph (x,noofiterations,convergence,f);

would use a declared function f which must have a single real parameter, while

> newtonraph (y,noofiterations,convergence,g);

would use a declared function *g* which must have only a single real parameter.

In this particular case we could also use a standard Pascal function such as $SIN(x)$ since its requirements match that of the parameter. Thus the following is also valid in this case:

> newtonraph (x,noofiterations,convergence,SIN);

We could also use procedures as parameters in a similar way. The only restriction is that the parameters of the procedures and functions called in this way must be value parameters.

8.6 Exercises

8.1. Starting from an initial value of 1.5, determine a root of the equation

$$\cos(x) - x/4 = 0$$

(a) by rewriting the equation as $x = x + \cos(x) - x/4$,
(b) by the Newton-Raphson method,
(c) by the secant method.

Note how many iterations were used in each case.

8.2. Write a program to determine a root of the equation

$$a - x^2 + bx^3 = 0 \quad \text{when } a = 56 \text{ and } b = 0.02$$

using the Newton-Raphson method.

For b taking values from 0.01 to 0.06 in steps of 0.01, produce a table of the values of the polynomial and its derivative for x taking values from -5 to 16 in steps of size 1.

Using these x values as starting values for the Newton-Raphson method, determine a root of the equation in each case. Summarise and explain what you observe. (*Note:* be careful at $x = 0$.)

8.3. Use and compare the methods described in this chapter to solve the equatio $f(x) = 0$ where $f(x) = 2 - 6x, x^2 - 1, x^3 - 64, 2\sin x - 1, \log x - 1$ and $\tan(x/4) - 1$, comparing both the time and the number of steps to conver-gence.

8.4. Calculate $5^{1/4}$, $7^{3/5}$, $\arcsin(1/3)$ using each of the methods. (*Hint:* think of $x^4 - 5$ etc.)

8.5. Use Newton's method to solve $5x - x^3 = 0$ from $x_0 = 1$, using the explicit expression for the derivative. What do you observe and why?

8.6. By sketching curves, determine how many solutions there are to the equation

$$5\cos x = 1 - x$$

and find them using the Newton-Raphson and secant methods.

9 Numerical Calculus

9.1 Numerical Differentiation

The derivative of a function $f(x)$ is defined in theory by

$$f'(x) = \lim_{h \to 0} \frac{f(x+h)-f(x)}{h}$$

Therefore we can obtain a numerical approximation using

$$\frac{f(x+h) - f(x)}{h} \text{ for small values of } h$$

The accuracy will clearly depend on the size of h. We can improve the approximation by letting h become small, but it is obvious we cannot let it become zero since we are dealing with numerical values. If h became zero we would be dividing by zero. However, before this becomes a problem the difference between $f(x)$ and $f(x+h)$ would become indistinguishable on the computer and this would lead to gross inaccuracies. Therefore, by reducing h in the classical theoretical manner, we can only achieve an accuracy within the limits of that allowed by the computer we are using.

This problem arises because computers store real numbers to only so many figures accuracy. While there may be a real difference between the numbers, we cannot see it numerically without resorting to special devices.

One way of overcoming this problem is by devising a more efficient way of estimating the quantity we require without reducing h as much. For example, it is more sensible to estimate the derivative using information on both sides of the value in which we are interested and estimate $f'(x)$ by

$$\frac{f(x+h) - f(x-h)}{2h}$$

Figure 9.1 illustrates the two methods and gives us an idea of the improvement that can be achieved.

If we consider the Taylor series expansion of both $f(x+h)$ and $f(x-h)$, we can obtain some measure of the improvement that can be expected. Now

$$f(x+h) = f(x) + h\,f'(x) + \frac{h^2}{2!}\,f''(x) + \frac{h^3}{3!}\,f'''(x) + \dots$$

106

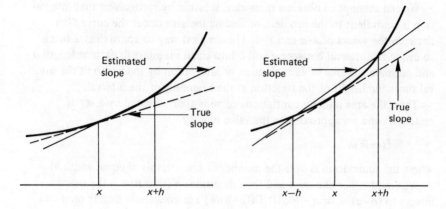

Figure 9.1

and $\qquad f(x-h) = f(x) - h\,f'(x) + \dfrac{h^2}{2!}\,f''(x) - \dfrac{h^3}{3!}\,f'''(x)\dots$

hence $\qquad \dfrac{f(x+h) - f(x)}{h} = f'(x) - \dfrac{h}{2!}\,f''(x) -$

and $\qquad \dfrac{f(x+h) - f(x-h)}{2h} = f'(x) + \dfrac{h^2}{6}\,f'''(x) +$

Thus, in the first case, the first term in the error is $h\,f''(x)/2$, while in the second case it is $h^2 f'''(x)/6$. Hence, since h is chosen to be small, the second error is potentially much smaller than the first and we have a much better approximation to the true derivative. We can ignore higher powers of h in this evaluation of the methods since the values of h considered are much less than 1.

9.2 Numerical Integration

Consider the problem of calculating

$$y = \int_a^b f(x)\,dx$$

where the function $f(x)$ is such that we cannot obtain an algebraic solution. For example

$$\int_a^b \exp(-x^2/2)\,dx \quad \text{or} \quad \int_0^1 x^{p-1}e^{-x}\,dx \quad \text{for } 0<p<1$$

We can attempt to obtain a numerical solution by recognising that integration is equivalent to the problem of finding the area under the curve f(x) between the values of $x=a$ and $x=b$. The simplest way to approximate this is to divide the interval between a and b into small subintervals each of length h, and estimate the area of each element of integration by the width of the interval times the height of the function at the beginning of the interval.

Thus, the area under a continuous curve is approximated by a set of rectangles and we approximate the value of y by

$$\Sigma \, f(a+ih)h$$

where the summation is over the number of subintervals of equal width h.

We could either choose h and then determine n (that is, n is the nearest integer to $(b-a)/h$, or $n = \mathrm{ROUND}((b-a)/h)$) or conversely decide on n and use it to determine h (that is, $h = (b-a)/n$). There is a preference for fixing n, since the h value used then covers the interval (a, b).

As is shown in figure 9.2, the effect of this is to use the shaded area as an estimate of the area under the curve.

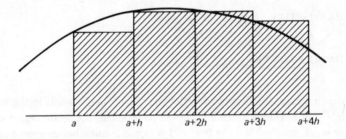

Figure 9.2

Clearly, the smaller the value of h, or correspondingly the larger the number of intervals that we use, the greater the accuracy that we can hope to achieve. However, this increases the amount of computation time. Thus we need to consider alternative methods which improve the accuracy without causing large increases in computation time.

From figure 9.2 we see that the error using this method is partly due to only using the height at the beginning of each interval to estimate the area. We should therefore immediately be able to improve the accuracy of estimation if we use either

(1) an estimate of the area of the rectangle using the height at the mid-point of the interval, or

(2) information at both ends of the interval and an estimation of the area of the element by the area of a trapezium rather than a rectangle.

Figure 9.3 illustrates this and we can see the immediate improvement that can be expected.

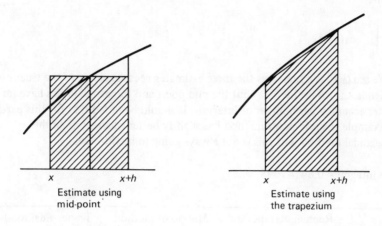

Figure 9.3

Using these ideas we produce

(1) the mid-point rule for numerical integration by which we estimate the integral by

$$h \sum_{i=0}^{n-1} f(a+(i+1/2)h) = h \ [f(a+h/2) + f(a+3h/2) + \ldots + f(a+(2n-1)h/2)]$$

and

(2) the trapezium rule for numerical integration, by which we estimate the integral by

$$\frac{h}{2} \sum_{i=1}^{n} [f(a+(i-1)h) + f(a+ih)]$$

$$= \frac{h}{2} \ [f(a) + 2f(a+h) + 2f(a+2h) + \ldots + 2f(a+(n-1)h) + f(b)]$$

where $h=(b-a)/n$ and n is the number of subintervals used.

For calculation purposes this can be written more efficiently as

$$h \ [f(a)/2 + f(a+h) + f(a+2h) + \ldots + f(a+(n-1)h) + f(b)/2]$$

Both of these methods can be programmed using the same order of operations as the basic rectangular method but achieve a far greater order of accuracy.

Example

The following example illustrates the result of applying all three methods to calculate

$$\int_0^1 \cos(x)\, dx$$

We see that as n increases the three estimates become closer to the true result (which is known in this case) but the mid-point and trapezoidal rules have far greater accuracy per number of intervals. It should be noted that for this particular example the mid-point method has slightly better accuracy than the trapezoidal method, but this is not always going to be the case.

True integral = 0.84147098.

No. of intervals	Rectangular method	Mid-point method	Trapezoidal method
10	0.86375453	0.84182170	0.84076964
100	0.84376246	0.84147449	0.84146397
1000	0.84170077	0.84147102	0.84147092
10000	0.84149395	0.84147098	0.84147097

9.2.1 Programming aspects

Since a finite number of terms is included in each estimate of the integral, the number being determined by the value of h and the interval (a, b), then any numerical integration method based on the above ideas is simply programmed by using a FOR loop.

The structure of these types of program is therefore:

(1) Determine the value of h, the interval size
or n, the number of terms.
(2) Set the integral to its initial value
for example in the case of the rectangular rule this is $f(a)$,
the ordinate at the beginning of the interval.
(3) Set x equal to the lower bound of the interval.
(4) Sum $n-1$ terms to the value of integral
increasing the value of x by h each time.
(5) Multiply the integral by the scaling factor h.

The summation stage is achieved by calculating the next value of x and adding the appropriate multiple of the function at this point.

Thus the rectangular estimate is calculated by adding $f(x)$ at each stage. In this case the scaling factor used in the last stage is h.

Hence the following Pascal code would estimate the integral using the rectangular rule.

```
h := (b - a) / n;
integral := f(a);
x := a;
FOR i := 1 TO n-1 DO
   BEGIN
   (* since we have already  included the function value
      corresponding  to the first value  of  x  we  must
      increase x before including the next term *)
      x := x + h;
      integral := integral + f(x);
   END;
integral := integral * h;
```

To program the mid-point rule we set the initial value in step (2) to $f(a + h/2)$ and the initial value of x in step (3) to $x + h/2$. The remainder of the steps are unchanged.

To program the trapezium rule we would set the initial value in step (2) to $(f(a) + f(b))/2$, the rest of the steps being unchanged.

The result would be the following in each case

Mid-point method

```
h := (b - a) / n;
integral := f(a + h / 2);
x := a + h / 2;
FOR i :=1 TO n-1 DO
    BEGIN
       x := x + h;
       integral := integral + f(x);
    END;
integral := integral * h;
```

Trapezoidal method

```
h := (b - a) / n;
integral := (f(a) + f(b)) / 2;
x := a;
FOR i := 1 TO n-1 DO
   BEGIN
      x := x + h;
      integral := integral + f(x);
   END;
   integral := integral * h;
```

In all three cases the function $f(x)$ would be supplied as a function earlier in the program.

The steps to perform the integral calculation should be placed in a procedure which would require the following parameters:

integral   a variable real parameter which will contain the estimate of
            the integration

a and *b*   two real value parameters giving the ends of the interval

n   an integer value parameter giving the number of intervals to
            be used

The remaining identifiers used, *h* , *x* and *i*, would be declared as local to the procedure, which would therefore be headed with the following statements:

```
PROCEDURE integrator (VAR integral : REAL;
                      a,b : REAL; n : INTEGER);
VAR h,x : REAL; i : INTEGER;
```

This structure forms the basis for any numerical integration program and in this form can easily be tested with a known integral before being applied to the unknown integral of interest.

9.3 Simpson's Rule

The three methods considered so far estimate the area under the curve over an interval of length h by multiplying the length of the interval by an estimate of the average height of the function over the interval. In the three methods these estimates are

rectangular method: height at the beginning of the interval

mid-point method: height at the mid-point of the interval

trapezoidal method: average of the height at the beginning and the
                     end of the interval

These methods use either a single point, which approximates the function over the interval by a line parallel to the x-axis, or two points, which means that we are approximating the function by a straight line.

To obtain further improvement in the estimation procedure we would expect to use three or more points, which means approximating the function over the interval by a polynomial. If we use a quadratic to approximate the function and estimate the average height of the quadratic by a weighted average of the ordinates at the end points and the mid-point, then another estimate of the area over an interval of length k is

$$\frac{k}{6} \ [f(x) + 4f(x+k/2) + f(x+k)]$$

which is known as Simpson's rule.

If we let $k = 2h$, the estimate of the integral over the interval $x, x+2h$ is given by

$$\frac{h}{3} \ [f(x) + 4f(x+h) + f(x+2h)]$$

If we now sum the results for the $n/2$ intervals of size $2h$ which cover the interval (a, b), the estimate of our integral becomes

$$\frac{h}{3} \ [f(a)+4f(a+h)+2f(a+2h)+4f(a+3h)+ \ldots +2f(a+(n-2)h)+4f(a+(n-1)h)+f(b)]$$

Note that n, the number of intervals used, must be even.

9.3.1 Programming aspects

Simpson's method can be programmed using the same structure as for our other three methods, making only the following three minor changes:

(1) In step (2) we set the initial value of the integral to $f(a)+f(b)$.
(2) In step (4) we add $4*f(x)$ to the value of the integral if the current term is odd and $2*f(x)$ if it is even. Here we can make use of the BOOLEAN function ODD(i) which has the value TRUE if i is odd and FALSE if i is even.
(3) In step 5 we multiply the integral by $h/3$ rather than h.

This leads us to the following:

```
h := (b - a) / n;
integral := f(a) + f(b);
x := a;
FOR i := 1 TO n-1 DO
    BEGIN
        x := x + h;
        IF ODD(i) THEN integral := integral + 4 * f(x)
                  ELSE integral := integral + 2 * f(x);
    END;
integral := integral * h / 3;
```

which can be used in the same procedure construction as before.

Example

Applying the method to the example considered in section 9.2.1, we get the following results:

True integral = 0.84147098

No. of intervals	Simpson's method
10	0.84147145
100	0.84147098
1000	0.84147099
10000	0.84147097

Comparing this with the results in section 9.2.1 we see that Simpson's method is as good using 100 intervals as the mid-point and trapezoidal methods were using 10 000 intervals. The answer is also accurate to five significant figures using only 10 intervals. Thus a considerable saving in computational effort is immediately obvious.

9.4   Errors in Numerical Integration

Both the trapezoidal rule and Simpson's rule can be constructed using a Taylor series expansion to approximate the function. Such a method of construction is more complicated mathematically than the heuristic arguments so far used to devise techniques for numerical integration. However, as well as giving a more theoretical justification, it also allows us to estimate the error involved in using such approximations. More importantly perhaps, it also allows us to compare techniques and evaluate their relative performance.

Using a Taylor series expansion we write

$$f(x+k) = f(x) + kf'(x) + \frac{k^2 f''(x)}{2!} + \dots$$

$$= a_0 + ka_1 + k^2 a_2 + \dots$$

where

$$a_m = \frac{1}{m!} \frac{d^m f(x)}{dx^m}$$

The trapezoidal rule is obtained by summing approximations to the integral over intervals of length h. Now

$$\int_x^{x+h} f(x) \, dx = \int_0^h f(x+k) \, dk$$

$$= \int_0^h [a_0 + ka_1 + k^2 a_2 + \dots] \, dk$$

$$= a_0 h + a_1 \frac{h^2}{2} + a_2 \frac{h^3}{2} + \dots$$

Also

$$f(x+h) + f(x) = 2a_0 + ha_1 + h^2 a_2 + \ldots$$

and

$$\frac{h}{2} \, [f(x+h) + f(x)] = a_0 h + a_1 \frac{h^2}{2} + a_2 \frac{h^3}{2} + \ldots$$

Therefore

$$\int_x^{x+h} f(x) \, dx = \frac{h}{2} \, [f(x+h) + f(x)] - a_2 \frac{h^3}{6} + \ldots$$

Hence if we assume that h is small enough so that terms involving powers of h that are greater than 2 can be considered negligible then

$$\int_x^{x+h} f(x) \, dx \quad \text{is approximately} \quad \frac{h}{2} \, [f(x+h)+f(x)]$$

which if we sum over all the intervals of this length we obtain the trapezoidal rule.

The term

$$-\frac{a_2 h^3}{6} = -\frac{h^3 f''(x)}{12}$$

gives us an estimate of the error involved in making such an approximation, since it will usually be the largest term in the part of the expansion that has been ignored in the truncation of the approximation to the integral. It is therefore usually called the *truncation error*.

Simpson's rule is obtained by considering intervals of length $2h$ and summing approximations to the integral over $x-h$ to $x+h$. Now

$$\int_{x-h}^{x+h} f(x) \, dx = \int_{-h}^{h} f(x+k) \, dk$$

$$= \int_{-h}^{h} [a_0 + a_1 k + a_2 k^2 + a_3 k^3 + a_4 k^4 + \ldots] \, dk$$

$$= 2ha_0 + 0 + 2a_2 \frac{h^3}{3} + 0 + 2a_4 \frac{h^5}{5} + \ldots$$

Notice here that all the odd terms in the sequence involving a disappear.

Now

$$f(x+h) + f(x-h) = 2a_0 + 2h^2 a_2 + 2a_4 h^4 + \ldots$$

$$\frac{h}{3}\,[f(x{+}h)+f(x{-}h)] + 4a_0\,\frac{h}{3} = 2a_0 h + 2a_2\,\frac{h^3}{3} + 2a_4\,\frac{h^5}{3} + \ldots$$

Hence

$$\int_{x-h}^{x+h} f(x)\,dx = \frac{h}{3}\,[f(x{-}h) + 4f(x) + f(x{+}h)] + \left[\frac{2a_4}{5} - \frac{2a_4}{3}\right] h^5 + \ldots$$

Hence if h is small enough for terms in h^5 to be considered negligible, the integral can be approximated by $[h/3]\,[f(x{-}h)+4f(x)+f(x{+}h)]$ which gives us Simpson's rule.

The term

$$-\frac{4a_4 h^5}{15} = \frac{4h^5}{15\times 4!}\,f''''(x) = -\frac{h^5}{90}\,f''''(x)$$

gives us an estimate of the error.

Comparing this with the truncation error for the trapezoidal rule, we immediately see why we observed such a dramatic improvement in the estimate of the integral obtained with different interval lengths. Using Simpson's rule we obtained as good an estimate using 100 intervals as that obtained using 10 000 intervals with the other methods.

9.5 Exercises

9.1. Write functions to calculate the derivative of $f(x)$ using the formula

$$\frac{f(x{+}h) - f(x)}{h}, \quad \frac{f(x{+}h) - f(x{-}h)}{2h}$$

Using $h = 10^{-n}$ ($n{=}1, 2 \ldots 10$), calculate $f(x)$ using each of the methods for a range of functions for which you know the answer. Compare the estimate with the true value.

9.2. The Gamma function is defined by the integral

$$\Gamma(r) = \int_0^1 x^{r-1} e^{-x}\,dx$$

Produce a table of values of $\Gamma(r)$ for $r = 0$ to 1 in steps of 0.1.

9.3. Estimate the value of the integral

$$\int_0^t 1/(1 - k^2 \sin^2 x)\,dx$$

for $k = 0.5$ with t taking values from 0 to 2π in steps of $\pi/2$.

9.4. Using a method of numerical integration, generate a table of values for the Normal distribution function given by

$$\Phi(x) = (1/2\pi) \int_{-\infty}^{x} \exp(-t^2/2)\, dt$$

for $x = 0$ to 2.5 in steps of 0.02. The integral should be accurate to at least four significant figures and can be checked against a set of normal probability tables. Display the table using the full width of your available printer. (*Hint:* use the symmetry of the Normal density to overcome the problem of the lower end of the interval being $-\infty$.)

9.5. Write a program to calculate $\int_0^1 \cos(\pi x/2)\, dx$ using the methods described in this chapter. For each method calculate the true error and the first term of the estimated error using the Taylor expansions described in the text, and comment on your results.

9.6. A method, not covered in the text, is that of Boole by which the integral $\int_a^b f(x)\, dx$ is approximated by

$$(h/8)\{f(a) + 3f(a+h) + 3f(a+2h) + 2f(a+3h) + \ldots + 3f(a+(n-1)h) + f(a+nh)\}$$

Write a procedure to estimate the integral using this method and then compare it with the results obtained by the other methods described in the text, on the integrals

$$\int_0^1 x^7\, dx \qquad \int_0^\pi \cos x\, dx \qquad \int_1^2 \log x\, dx \qquad \int_0^1 \sqrt{x}\, dx$$

$$\int_1^2 \sqrt{x}\, dx \qquad \int_{10}^{11} \sqrt{x}\, dx \qquad \int_{0.01}^1 x^{-1/2}\, dx$$

Try a variety of different values of n (for example, 10, 50, 100) where n is the number of intervals used, and summarise your conclusions, commenting on the results.

9.7. Use your programs in exercise **9.6** to estimate

$$\int_0^1 4/(1+x^2)\, dx \quad \text{and} \quad d\int_0^1 \sqrt{(1-x^2)}\, dx$$

Which is better in each case?

10 First-order Differential Equations

10.1 Introduction

One of the common problems in applied mathematics is to solve differential equations of the form

$$\frac{dy}{dx} = f(x,y)$$

given an initial value of y_0 at the point x_0.

There are many instances when there is no known closed algebraic form available as a solution of the problem. For example

$$\frac{dy}{dx} = x^2 + y^2 \quad \text{or} \quad \frac{dy}{dx} = \exp(kx^2)$$

In these cases we need to obtain a numerical solution to the problem in the form of a table of values of y at a sequence of equally spaced x values, namely $x_0, x_0+h, x_0+2h, x_0+3h \ldots$.

From these we will be able to draw a graph illustrating an approximation to the unknown function y from that particular starting value.

10.2 Euler's method

We begin by considering the simplest method which is achieved by approximating the curve over each x interval by a straight line using the current estimate of the gradient of the curve. From figure 10.1 we see that, given the initial point (x_0, y_0) and the gradient $f(x_0, y_0)$, then the curve over the interval (x_0, x_0+h) can be approximated by the straight line

$$y = y_0 + (x-x_0)\, f(x_0, y_0)$$

giving us the value of y at $x_0 + h$ as

$$y_1 = y_0 + h\, f(x_0, y_0)$$

118

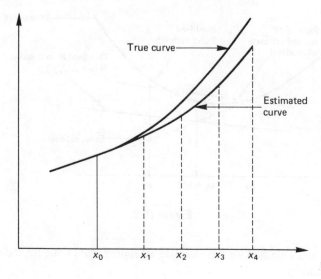

Figure 10.1

Once we have this estimate of the next point on the curve, we can use a similar extrapolation over the next interval (x_0+h, x_0+2h) to give us

$$y_2 = y_1 + h\, f(x_1, y_1) \quad \text{where } x_1 = x_0 + h$$

Proceeding in this way, we generate a sequence of approximations to the unknown curve using the sequential mapping

$$y_{n+1} = y_n + h\, f(x_n, y_n) \quad \text{where } x_n = x_0 + nh$$

The accuracy of this approximation obviously depends on the magnitude of h — that is, the size of step taken in the x direction — and will improve as the value of h becomes small although, as with our other applications of numerical methods, we cannot reduce h beyond a certain point.

Since the estimate at each stage is an approximation based on earlier approximations, any errors involved will accumulate so that the further from our starting value the less accurate our estimate becomes. Hence it is crucial to produce as accurate an estimate as possible.

10.3 Modified Euler

We can improve Euler's method by attempting to find a more accurate estimate of the gradient of the line from (x_0, y_0) to (x_1, y_1). Figure 10.2 shows that a more accurate estimate of the curve is obtained by using the gradient at the mid-point of the interval.

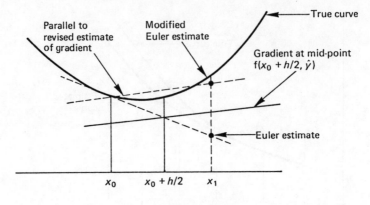

Figure 10.2

Using Euler's method we can calculate an approximate value for the curve at the mid-point by

$$y(x_0 + h/2) = y_0 + h\,f(x_0, y_0)/2$$

Using this, we can calculate an estimate of the gradient of the curve at that point, that is

$$g = f(x_0 + h/2, y_0 + h\,f(x_0, y_0)/2)$$

and using this estimate of the gradient we can estimate the value of the curve at $x_0 + h$ using the equation

$$y_1 = y_0 + h\,g$$

Repeating this process, we have another sequential mapping that will produce a series of estimates to the curve.

10.4 Example

Consider the initial value problem

$$f(x, y) = x + y \text{ where } y_0 = 1 \text{ at } x_0 = 0$$

In this case we can use the exact solution, namely

$$y = 2e^x - x - 1$$

to allow us to assess how the two methods are performing. Tables 10.1 and 10.2 show the results using step sizes of $h = 0.10$ and $h = 0.05$ respectively. The error quoted is the percentage relative error = absolute error/true result $\times$ 100.

Table 10.1 $h = 0.10$

x	Exact	Euler	% error	Modified	% error
0.10	1.11034	1.10000	0.93141	1.11000	0.03079
0.20	1.24281	1.22000	1.83500	1.24205	0.06079
0.30	1.39972	1.36200	2.69466	1.39847	0.08947
0.40	1.58365	1.52820	3.50137	1.58180	0.11652
0.50	1.79744	1.72102	4.25174	1.79489	0.14181
0.60	2.04424	1.94312	4.94637	2.04086	0.16535
0.70	2.32751	2.19743	5.58844	2.32315	0.18724
0.80	2.65108	2.48718	6.18254	2.64558	0.20761
0.90	3.01921	2.81590	6.73392	3.01236	0.22664
1.00	3.43656	3.18748	7.24790	3.42816	0.24449

Table 10.2 $h = 0.05$

x	Exact	Euler	% error	Modified	% error
0.05	1.05254	1.05000	0.24153	1.05250	0.00401
0.10	1.11034	1.10500	0.48110	1.11025	0.00799
0.15	1.17367	1.16525	0.71728	1.17353	0.01192
0.20	1.24281	1.23101	0.94890	1.24261	0.01578
0.25	1.31805	1.30256	1.17505	1.31779	0.01955
0.30	1.39972	1.38019	1.39502	1.39939	0.02322
0.35	1.48814	1.46420	1.60834	1.48774	0.02679
0.40	1.58365	1.55491	1.81470	1.58317	0.03024
0.45	1.68662	1.65266	2.01396	1.68606	0.03358
0.50	1.79744	1.75779	2.20609	1.79678	0.03681
0.55	1.91651	1.87068	2.39119	1.91574	0.03992
0.60	2.04424	1.99171	2.56942	2.04336	0.04292
0.65	2.18108	2.12130	2.74100	2.18008	0.04582
0.70	2.32751	2.25986	2.90621	2.32637	0.04861
0.75	2.48400	2.40786	3.06537	2.48273	0.05130
0.80	2.65108	2.56575	3.21879	2.64965	0.05390
0.85	2.82929	2.73404	3.36681	2.82770	0.05641
0.90	3.01921	2.91324	3.50979	3.01743	0.05884
0.95	3.22142	3.10390	3.64805	3.21945	0.06120
1.00	3.43656	3.30660	3.78192	3.43438	0.06348

We see two points from these results:

(1) The smaller the step size the smaller the accumulated error, but this obviously increases the amount of computation involved since halving the step size means doubling the function evaluations we need to cover the same interval.

(2) The modified Euler is vastly superior to Euler's method, thereby offsetting the extra amount of computing it involves.

10.5 Programming Aspects

The above methods are examples of a family of methods (called the Runge–Kutta methods) which are based on the following two principles:

(a) only information about the current point is used to calculate the next point and

(b) no derivatives of $f(x, y)$ are used.

They can all be programmed using the following structure

```
┌─────────────────┐        ┌─────────────────────────┐
│ Enter           │        │ Estimate the curve      │
│ initial values  │        │ over a fixed number of  │
│ number of steps │        │ steps                   │
│ and step size   │        │                         │
└─────────────────┘        └─────────────────────────┘
```

It is usual to estimate over a fixed number of steps, n, which, using the step size, h, is equivalent to estimating the curve over the interval $(x_0, x_0 + nh)$ on the x-axis.

The estimation stage is therefore usually programmed as a FOR loop containing the following steps:

> calculate the next value of y
> increment the x value
> output the values of x and y

The first of these three steps depends on the method chosen and can be one or more statements that perform the sequential mapping from y_n to y_{n+1}.

Thus, for Euler's method the loop would consist of the following:

```
FOR i:= 1 TO n DO
    BEGIN
        yapprox := yapprox + h * f(x,yapprox);
        x := x + h;
        WRITELN(x:6:2,yapprox:16:10);
    END;
```

The function f is the gradient function which must be defined at the top of the program.

The modified Euler method requires the first line in the BEGIN ... END block to be replaced by either the pair of lines

```
ymid := yapprox + h * f(x,yapprox) / 2;
yapprox := yapprox + h * f(x + h / 2,ymid);
```

or the single line

```
yapprox := yapprox + h * f(x + h / 2,
              yapprox + h * f(x,yapprox) / 2);
```

The above recipe can be used to handle any of the methods in the Runge-Kutta family.

10.6 Predictor–corrector Methods

In the modified Euler method we require evaluation of our function $f(x, y)$ for at least two points and only use the most recent value to estimate the next. However, once we have started the solution process we will have a set of values which we could use as prior values to predict the value of y_{n+1}. We could, for example, use the modified Euler method over the interval (x_{n-1}, x_{n+1}) with x_n as the mid-point, giving

$$y_{n+1} = y_{n-1} + 2hf(x_n, y_n)$$

as a prediction of the next value. Once an approximate value for y_{n+1} has been obtained, a more accurate estimate can be calculated as follows

(a) Calculate $y'_{n+1} = f(x_{n+1}, y_{n+1})$.
(b) Draw a line through the point (x_n, y_n) with a gradient that is the average of y'_n and y'_{n+1}.
(c) Take the y value on this line at x_{n+1} as a new estimate of y_{n+1}. The new value for y_{n+1} is given by

$$y_{n+1} = y_n + \frac{h}{2} \left(f(x_n, y_n) + f(x_{n+1}, y_{n+1}) \right)$$

This process is called the correction process since it corrects the predicted value using new information about the gradient. The combined process is an example of a predictor–corrector method. (Note that only one function evaluation is necessary at each step, rather than the two used in the modified Euler method.)

The initial step of estimating y_{n+1} given the previous set of values is called the prediction process and, since it requires more than the current estimate to predict the value of y_{n+1}, we must use another method such as the modified Euler method to start the process.

The basic structure of the problem from which we can produce a program is as follows:

(1) Use modified Euler to obtain
first estimate at $x+h$.
(2) Increasing the x value by h each time
perform a prediction step
perform a correction step
Until the x value reaches the end of the interval.

We illustrate this in the code below in which all the variables are REAL. The function f will contain $f(x, y)$, the derivative of y with respect to x. If the exact solution is known, using a second function *exactf(x)* allows us to calculate the percentage error so that we can assess the performance of the method as shown in the program.

```
BEGIN
(* enter the initial values for x and ycurrent  and the
final value for x *)
  WRITELN('enter the initial values for x and y');
  READ(x, ycurrent);
  WRITELN('What is the step length');
  READ(h);
  WRITELN('What is the final x value');
  READ(finalx);

(* set up the table heading *)
  WRITELN('x':6,'exact':11,'estimated':11,
          '%error':11,'corrections':13);

(* use modeuler to calculate the first estimate in yest
*)
  ymid := ycurrent + h * f(x,ycurrent) / 2;
  yyest := ycurrent + h * f(x + h / 2, ymid);

(* increment x  and  display  the estimate of y at this
point *)
  x := x + h;
  nextgrad:=f(x,yest);
  yerr := ABS(yest - exactf(x)) * 100 / exactf(x);
  WRITELN(x:6:2, exactf(x):11:5, yest:11:5, yerr:11:5);
(* Apply  the  predictor/corrector  method   until the
    value of x reaches the end of the required interval
*)
```

```
    REPEAT
(*    move the estimates of y  back  one  time point and
      store the current estimate of the gradient *)
      yprevious := ycurrent;
      ycurrent := yest;
      ygrad:=nextgrad;

(*  perform the prediction step *)
      yest:=yprevious + 2 * h * ygrad;

(*  perform the correction step *)
      nextgrad:=f(x+h,yest);
      yest:=ycurrent+(ygrad+nextgrad) * h / 2;

(* increment x and display the value of the estimate of
    y at this point *)
      x := x + h;
      yerr := ABS(yest - exactf(x)) * 100 / exactf(x);
      WRITELN(x:6:2, exactf(x):11:5, yest:11:5, yerr:11:5);
    UNTIL x >= finalx;
END.
```

Note that the value of x is moved forward before we calculate the true value and the error, since the estimate we have calculated is one of the value at $x+h$ using the information at x. Any comparison should therefore be made with the true function at the next point and not the current point.

If we apply this method to the example considered in section 10.4, we get the results shown in table 10.3 using a step size of 0.1.

Table 10.3   Results using a predictor–corrector method for $h = 0.1$

x	Exact	Estimated	% error
0.10	1.11034	1.11000	0.03079
0.20	1.24281	1.24260	0.01654
0.30	1.39972	1.39962	0.00697
0.40	1.58365	1.58365	0.00029
0.50	1.79744	1.79757	0.00719
0.60	2.04424	2.04452	0.01377
0.70	2.32751	2.32797	0.01995
0.80	2.65108	2.65176	0.02575
0.90	3.01921	3.02015	0.03119
1.00	3.43656	3.43781	0.03630

It should be noted that while we see a significant improvement in this case, it is not always the case that the P–C method outperforms the Runge–Kutta method.

10.7 Comment

The Runge-Kutta methods such as the Euler and modified Euler are easy to program since they require only a single value to start the calculation. The predictor–corrector methods, however, require two or more initial values to start up the calculation and the accuracy of subsequent values clearly depends on these initial values. In general, P–C methods require fewer function evaluations than Runge-Kutta methods of the same order. Thus, while they produce comparable results the predictor–corrector methods are generally quicker.

It should be noted that, on its own, the predictor step would not be accurate enough since it contains no new information based on the current point. We need to correct the prediction using an up-to-date estimate of the gradient. A further possibility is to iterate the correction step until no further improvement in the corrected value can be achieved. However, this cannot achieve accuracy beyond the order of the correction step and a better way of using the extra computation to improve the accuracy is to reduce the step size.

Stability of the solution is another feature that can be used to assess the relative performance of the methods and help us to choose an appropriate method. Details of these and other problems, together with descriptions of higher-order systems, can be found in most texts on numerical analysis. See, for example, Phillips and Taylor (1973). A more detailed description, together with practical comments, is given in Maron (1987).

10.8 Exercises

10.1. Write a program to solve the differential equation

$$\frac{dy}{dx} = y$$

from initial value $x=0$, $y=1.0$ using the Euler and modified Euler methods. Perform the calculations over the interval $x = 0$ to $x = 1$ with step size h reducing from $h = 0.1$ to 0.001 by factors of 10. Display only the estimates at $x = 0, 0.2, 0.4, 0.6, 0.8$ and 1. Compare your results with the exact solution.

10.2. Show that $y = 1/(a - \cos x^2)$ is the solution of the differential equation

$$\frac{dy}{dx} = -2xy^2 \sin x^2$$

Starting with the initial condition $x = 0$, $y = 5$, determine the numerical solution at $x = 1$ and $x = 2$ using Euler's method and the modified Euler method using step sizes h of 0.1 and 0.01. Compare the results with the true results and comment on what you observe.

10.3. The initial value problem

$$\frac{dy}{dx} = 2x + y \quad y(0) = 1$$

has analytic solution $y(x) = 3e^x - 2x - 2$.

Using $h = 0.1$ and $h = 0.01$, write programs to solve this problem using the Euler, modified Euler and the predictor–corrector methods. Display your results, together with the correct answer, to six significant figures, in a table for x values 0, (0.1), 1.0. Comment on your results.

10.4. The initial value problem for the vertical motion of a projectile in non-dimensional form is

$$\frac{d}{dz}(w^2/2) + (kr_0)w^2 = -1/z^2$$

where $w = r_0 v_0^2/gm$ at $z = 1$.

By using the Euler and the predictor–corrector methods to produce approximate solutions to the differential equation for $w > 0$, tabulate and draw a graph of w in terms of z for $kr_0 = 1/4$ and $1/1000$ respectively.

Use $r_0 v_0^2/gm = 1, 2$ and 4 for each value of kr_0.

How would you proceed for $w < 0$?

10.5. The Euler and modified Euler methods described in this chapter are examples of first-order and second-order Runga-Kutta methods. A third-order Runge-Kutta method is given by

$$k_1 = hf(x_0, y_0)$$
$$k_2 = hf(x_0 + h/2, y_0 + k_1/2)$$
$$k_3 = hf(x_0 + h, y_0 - k_1 + 2k_2)$$
$$y_1 = y_0 + (k_1 + 4k_2 + k_3)/6$$

while a fourth-order method is given by

$$k_1 = hf(x_0, y_0)$$
$$k_2 = hf(x_0 + h/2, y_0 + k_1/2)$$
$$k_3 = hf(x_0 + h/2, y_0 + k_2/2)$$
$$k_4 = hf(x_0 + h, y_0 + k_3)$$
$$y_1 = y_0 + (k_1 + 2k_2 + 2k_3 + k_4)/6$$

Write programs for both these methods and then use them on the problems in exercises **10.1, 10.2** and **10.3** above, comparing the results with the methods used.

10.6. A fourth-order predictor-corrector method is Adams-Bashforth-Moulton which uses the following:

Prediction step

$$y_{n+1} = y_n + h(55y'_n - 59y'_{n-1} + 37y'_{n-2} - 9y'_{n-3})/24$$

Correction step

$$y_{n+1} = y_n + h(5y'_{n+1} + 8y'_n - y'_{n-1})/12$$

Using the fourth-order Runga–Kutta method in exercise **10.5** to estimate the first three values, write a program for this method and apply it to the problems given above, comparing the results with the other methods.

11 Arrays

11.1 Vectors or One-dimensional Arrays

The examples we have discussed so far have all involved a few named variables
and the input of only one or two values into the program. This has not proved
a problem in these examples but such an approach would become cumbersome
very rapidly when we consider problems involving vectors and matrices. Data
analysis becomes practically impossible unless there is some way of storing and
manipulating whole blocks of numerical values.

A vector of values can be stored in an array which allocates a number of
storage locations to a single name, the individual values being identified by a
subscript. Following the normal Pascal procedure, we define an array before
using it. In specifying the array we must indicate the range of subscripts allowed
and the type of values which can be assigned to the array. The definition is done
as part of the VAR statement and an array is defined as follows

> identifiername : ARRAY [n1 . . n2] OF type

where identifiername is the user-chosen name of the array, n1 is the bottom end
of the range of subscripts allowed and n2 is the top of the range of subscripts
allowed, and type may be any of the allowed types. The range of subscripts is
separated by two dots as illustrated, and the square brackets are used to
indicate that the subscripts are contained within them.

Examples

x : ARRAY [1. .10] OF REAL
> defines an array x which can contain 10 real values
> the subscripts are in the range 1 to 10

I : ARRAY [16. .145] OF INTEGER
> defines an array I which can contain 130 integer values
> the subscripts are in the range 16 to 145

Y : ARRAY [−5. .5] OF REAL
> defines an array Y which can contain 11 real values
> the subscripts are in the range −5 to 5

We illustrate the use of arrays in the following version of the program to calculate the mean of a sample of *n* values and then calculate the deviation of each observation from the mean, storing the result in a second array y. Using our earlier program where we read each sample value into a single identifier and added it to the sum before reading the next term, we would not be able to perform this extra operation.

```
PROGRAM firstarrayexample(INPUT,OUTPUT);

VAR mean : REAL;
    i,n : INTEGER;
    x,y : ARRAY [1..10] OF REAL;

BEGIN

(* read the data into the array x *)
   READ(n);
   FOR i := 1 to n DO
      READ(x[i]);

(* calculation of the mean *)
   mean := 0;
   FOR i := 1 TO n DO
      mean := mean + x[i];
   mean := mean / n;

(* display the mean *)
   WRITELN('mean is ',mean:6:2);

(* calculate the deviations from the mean
   and display them *)
   FOR i := 1 TO n DO
      BEGIN
         y[i] := x[i] - mean;
         WRITE(y[i])
      END;

END.
```

Each part of the array has two values associated with it:

(a) the subscript and
(b) the actual value assigned to it.

The value of the subscript used when values of the array are required must be within the specified range or an error occurs at the time the program is run. The values in an array can be accessed randomly as well as in order (as was the case in the above example), and the subscript can be an expression provided the result is in the allowed range and of the correct type.

11.1.1 Example

The Greek mathematician Eratosthenes proposed the following to determine the prime numbers within the list of positive integers greater than 1.

The first number in the list is removed and called a prime. All the numbers in the list that are multiples of this number are then deleted from the list. The smallest integer left in the list is then selected as the next prime and the process repeated until no numbers remain in the list.

This process is known as the sieve of Eratosthenes and, while it cannot be used to determine large primes because of the time involved, it is a useful way of listing the lower-order primes. It is also a useful introductory example in the use of arrays.

The neatest way to handle this on a computer is to store the numbers 2 to n in an array of type INTEGER. Thus the array will be declared as x : ARRAY[2 .. n] OF INTEGER where n is the largest number in the list. Then as a number is identified as being a multiple of a prime, the value in that position of the array is replaced by zero to signify that it is no longer in the list. Once all the primes have been identified, we simply display the non-zero terms in the array.

We therefore proceed as follows:

1. FOR i := 2 to n
        assign the value i to x[i]
2. FOR each value in the array in turn
            IF it has not been deleted — that is, set to zero — THEN
                3.   delete all multiples of it by setting them to zero
4. Display the result.

Step 3 consists of taking each value in the array to the right of the latest prime and, if that value is non-zero, testing whether it is a multiple of the latest prime. If it is identified as a multiple of the prime, set its value to zero.

An integer n is a multiple of an integer m if the remainder from dividing n by m is zero. That is if n MOD $m = 0$.

Thus, if the current prime is in position i, step 3 becomes

```
FOR j:= i + 1 to n
   IF x[j] is non-zero THEN
      IF x[j] MOD x[i] is zero THEN
                        set x[j] to zero
```

If we now piece together the whole algorithm we get the following:

```
(* step 1 *)
FOR i := 2 TO n DO
    x[i] := i;
(* step 2 *)
FOR i := 2 TO n DO
    IF x[i] <> 0 THEN
    (* step 3 *)
    FOR j := i + 1 TO n DO
        IF x[j] <> 0 THEN
            IF x[j] MOD x[i] = 0 THEN x[j] := 0;

(* step 4
    the numbers are displayed 16 to a  line  using 4
    positions    for   each   number   and   a   space   in
    between.      Using   the   method   described   in
    chapter 6 this covers an 80 character screen *)
k := 0;
FOR i := 2 TO n DO
    IF x[i] <> 0 THEN
        BEGIN
            WRITE(x[i]:4, ' ');
            k := k + 1;
            IF k = 16 THEN
                BEGIN
                    k := 0;
                    WRITELN;
                END;
        END;
```

When we write an algorithm we should step back and check whether it is written in the most efficient way. Knowledge of mathematical results, common sense and the way in which calculations are performed can often result in a more efficient use of an algorithm. This particular case is an example in which we are performing redundant operations through not applying a little mathematical knowledge. This stems from the simple observation that if a number has prime factors then at least one of these must be less than $\sqrt{n}$. Thus step 2 can be modified in the sense that we only need to check for multiples of primes up to the nearest integer below $\sqrt{n}$ rather than all terms up to n. This is because once we have removed all the terms in the list that are multiples of primes up to $\sqrt{n}$, then the remaining values must all be prime. Hence any subsequent testing is unnecessary. We can therefore improve our algorithm by replacing the line FOR i := 2 TO n DO at the beginning of step 2 by the lines

```
m := TRUNC(SQRT(n));
FOR i := 2 TO m DO
```

11.2 Specification of Subscript Ranges

We have seen that the range of subscripts can be specified by supplying the two numerical values for the limits of the range. We can, however, also use identifiers to indicate the limits of the range, provided their values have been set as constants using the CONST statement.

The advantage of this is that the size of the arrays, particularly if there are many of them with the same dimension, can be easily changed by simply editing the CONST statement. It also means that the minimum and maximum subscript values are available for use in the program. They could, for example, be used to protect the values of subscripts used in the program in such a way that the array variable was not used if the intended subscript was out of bounds.

The example program we considered earlier would be better written if the array was defined using two statements to define it as follows:

```
CONST lowbd = 1; upperbd = 10;

VAR x : ARRAY [lowbd..upperbd] OF REAL;
```

11.3 Multi-dimensional Arrays

There are instances when we want to work in terms of many dimensions. Obvious examples are matrices in two dimensions or tables of values which can take on any number of dimensions. A two-dimensional array or matrix can be thought of as an array of arrays or more simply as an array with two subscripts indicating row and column. There are two ways of defining such quantities:

either   identifiername : ARRAY[n1. .n2] OF ARRAY[n3 . .n4] OF type;

or   identifiername : ARRAY[n1. .n2,n3. .n4] OF type;

where n1 to n2 indicates the range of the first subscript which represents the row of the matrix, and n3 to n4 indicates the range of the second subscript which indicates the column of the matrix. We can also specify the range limits using constants specified earlier in the program, just as we can for one-dimensional arrays.

Once the arrays have been defined they can be used in the same way as any other identifier provided the correct number of subscripts are used and they are all within their specified ranges.

11.3.1 Examples

The next set of examples illustrates how information can be read into matrices
and the sum and product of two matrices can be calculated.

```
FOR i :=1 TO n DO
   FOR j :=1 TO m  DO
       READ(matrix[i,j]);

FOR i :=1 TO n DO
   FOR j :=1 TO m DO
       matsum[i,j] := mata[i,j] + matb[i,j);

FOR i := 1 TO n DO
   FOR j := 1 TO p DO
     BEGIN
       matprod[i,j] := 0;
       FOR k := 1 TO m DO
          matprod[i,j] := matprod[i,j] +
                               mata[i,k] * matb[k,j];
     END;
```

11.4 Passing Arrays to Procedures

Arrays can be used as global identifiers in the same way as the other types of
identifiers that we have previously met. However, if we want to pass them as
parameters we must declare them in a special way. We must associate a user-
defined type name with the array and then use this type name in the definition
of the variable both in the main program and the procedure statement. We can
define the type using the command

```
TYPE name = ARRAY[n1..n2] OF type;
```

Then, whenever we want to define an array with that range of subscripts, we
simply use the name as a type in the variable definition list. For example

```
CONST n1 = 1; n2 = 100; n3 = 50;
TYPE vector = ARRAY[n1..n2] OF REAL;
     matrix = ARRAY[n1..n2,n1..n3] OF REAL;
VAR x,y : vector;
    mata,matb : matrix;
```

sets up two user-declared types. These are *vector*, a one-dimensional array with
subscripts in the range 1 to 100, and *matrix*, a two-dimensional array with subscripts 1 to 100 by 1 to 50. It then defines four variables, x and y to be of type
vector — that is, one-dimensional arrays with subscripts range 1 to 100 — and

mata and *matb* to be of type *matrix* — that is, to be used as 100 by 50 matrices or two-dimensional arrays.

We can also use the declared types in the procedure statement. For example

```
PROCEDURE matread(VAR mat:matrix;rows,columns:INTEGER);
```

declares a procedure with three parameters, the first of which is a variable parameter with type *matrix*, the type having been declared earlier in the program. The other two parameters are value parameters having the standard type INTEGER.

11.4.1 Example

We can illustrate the use of two-dimensional arrays and procedures by considering the problem of reading in two matrices, multiplying them and printing out the result.

The basic structure of the problem consists of four steps which can be fulfilled by using three procedures:

Step 1. Read in matrix *A*.
Step 2. Read in matrix *B*.
Step 3. Multiply the matrices and store the result in *C*.
Step 4. Write out the result stored in *C*.

The three procedures that we need to write are:

matread	to read in the matrices
matmul	to perform the multiplication and
displaymat	to output the matrix

We therefore need to declare three matrices, each of which has two integer values associated with it, which record the number of rows and columns being used in the matrix.

The next problem is to declare the dimensions of the matrices, in terms of storage locations that are going to be allowed in the computer's memory. It is useful to make the program as general as possible. The dimensions of matrix *C* are determined by the number of rows in *A* and the number of columns in *B*, hence to avoid declaring too many matrices we could declare a single type of square matrix so that the program can handle all matrices up to the maximum stated dimension which equals the maximum required for either of the matrices *A* or *B*.

This is inefficient in the use of the computer's memory for any particular problem, but it does make the program easier to write and means that the procedures are easily used in other programs.

Recall that matrix multiplication is only possible if the number of columns in the first matrix is equal to the number of rows in the second. Thus, we can safeguard our basic program by modifying the structure as follows using a BOOLEAN variable set in procedure *matmul* to indicate if the multiplication is allowed:

Step 1.   Read in matrix *A* and its number of rows and columns.

Step 2.   Read in matrix *B* and its number of rows and columns.

Step 3.   IF number of columns of *A* equals the number of rows of *B*
  THEN
               multiply the matrices storing the result in *C*
               and set a Boolean variable equal to true
  ELSE
               output a message that the matrices are incompatible for
               multiplication and set the Boolean variable to false.

Step 4.   IF the matrix multiplication was allowed
  THEN output the result stored in matrix *C*.

If we now write the program along these lines the result is the program given below.

```
PROGRAM matrixmultiplication (datafile,resultsfile);

(*  Program  which  reads  two  matrices  from  a  file
    datafile and outputs the result of multiplying them
    to a file resultsfile *)

CONST maxd = 10;
TYPE matrix = ARRAY[1..maxd,1..maxd] OF REAL;
VAR mata,matb,matc : matrix;
  rowa,rowb,rowc,cola,colb,colc : INTEGER;
  datafile,resultsfile : TEXT;
  mulpos : BOOLEAN;

PROCEDURE matread(VAR row,col : INTEGER;
                      VAR mat : matrix);
(* procedure for reading a  matrix  to be stored in mat
   with row and column information  stored  in  row and
   col respectively *)

  VAR i,j : INTEGER;

  BEGIN
    READ(datafile,row,col);
    FOR i :=1 TO row DO
      FOR j :=1 TO col DO
        READ(datafile,mat[i,j]);
  END;
```

```
PROCEDURE matmul(rowa,cola,rowb,colb : INTEGER;
                 VAR mata,matb,matc:matrix;
                 VAR rowc,colc : INTEGER;
                 VAR mulpos : BOOLEAN);
(* procedure to multiply matrices mata and matb storing
   the results in matc *)

   VAR i,j,k : INTEGER;

   BEGIN

     IF cola = rowb THEN
        BEGIN
          mulpos := TRUE;
          (* set up the dimensions of matrix C *)
          rowc := rowa;
          colc := colb;
          (* perform matrix multiplication *)
          FOR i := 1 TO rowc DO
            FOR j := 1 TO colc DO
              BEGIN
                matc[i,j] :=0.0;
                FOR k :=1 TO cola DO
                  matc[i,j] := matc[i,j] +
                                    mata[i,k] * matb[k,j];
              END;
        END

     ELSE

        BEGIN
          mulpos := FALSE;
          WRITELN('matrix multiplication impossible');
        END;

   END;


PROCEDURE displaymat(row,col : INTEGER;
                       VAR mat : matrix);
(* procedure for displaying a matrix *)

   VAR i,j : INTEGER;

   BEGIN
     FOR i :=1 TO row DO
        BEGIN
          FOR j :=1 TO col DO
            WRITE(resultsfile,mat[i,j]:7:1);
(* note that this next  line is necessary to  terminate
   the display of each line of the matrix *)
          WRITELN(resultsfile);
        END;
   END;
```

```
(* the main part of the program is as follows *)
BEGIN

(* set up the files for entry and receipt of results *)
  RESET(datafile);
  REWRITE(resultsfile);

(* steps 1 and 2 read in the two matrices *)
  matread(rowa,cola,mata);
  matread(rowb,colb,matb);

(* perform the multiplication of the matrices *)
  matmul(rowa,cola,rowb,colb,mata,matb,
                            matc,rowc,colc,mulpos);

(* If the multiplication is  allowed output the results
*)
  IF mulpos THEN
  BEGIN
    WRITELN(resultsfile,
                'The result of multiplying the matrix');
    WRITELN(resultsfile);
    displaymat(rowa,cola,mata);
    WRITELN(resultsfile);
    WRITELN(resultsfile,'by the matrix');
    WRITELN(resultsfile);
    displaymat(rowb,colb,matb);
    WRITELN(resultsfile);
    WRITELN(resultsfile,' is the matrix');
    WRITELN(resultsfile);
    displaymat(rowc,colc,matc);
  END;

END.
```

The result of running this program with the datafile

```
3 4
1 2 3 4
5 6 7 8
9 10 11 12
4 5
1 2 3 4 5
6 7 8 9 10
11 12 13 14 15
16 17 18 19 20
```

is a result's file containing the following:

The result of multiplying the matrix

```
1.0   2.0   3.0   4.0
5.0   6.0   7.0   8.0
9.0  10.0  11.0  12.0
```

by the matrix

```
 1.0   2.0   3.0   4.0   5.0
 6.0   7.0   8.0   9.0  10.0
11.0  12.0  13.0  14.0  15.0
16.0  17.0  18.0  19.0  20.0
```

is the matrix

```
110.0  120.0  130.0  140.0  150.0
246.0  272.0  298.0  324.0  350.0
382.0  424.0  466.0  508.0  550.0
```

11.5 Roots of a Polynomial

We conclude this chapter with an example which combines the use of arrays
with the techniques of iteration that we considered in chapter 8.

Consider the polynomial

$$f(x) = a_0 x^n + a_1 x^{n-1} + \ldots + a_{n-1} x + a_n$$

In chapter 3 we suggested that an alternative method for evaluating the
second-order polynomial $ax^2 + bx + c$ was to use $(ax + b)x + c$. When we consider the problem of evaluating the nth-order polynomial $f(x)$, this technique
not only makes the calculation more efficient but makes it easier to program
the general case. Extending the method introduced in chapter 3, we calculate
the value of $f(x)$ as follows:

$$\left.\begin{aligned}
f &= a_0 \\
f &= fx + a_1 \\
f &= fx + a_2 \\
&\;\;\vdots \\
f &= fx + a_{n-1} \\
f &= fx + a_n
\end{aligned}\right\} \tag{11.1}$$

If the coefficients of the polynomial are stored in an array a[0. .n] of type
REAL, then we can perform this sequence of operations using

```
f := a[0];
FOR i := 1 TO n DO
    f := f * x + a[i];
```

To find the root of this polynomial using the Newton–Raphson method we also need the derivative

$$f'(x) = na_0 x^{n-1} + (n-1)a_1 x^{n-2} + \ldots + 2a_{n-2} x + a_{n-1}$$

This can be calculated in the same way as the original polynomial using

$$g = na_0$$
$$g = gx + (n-1)a_1$$
.
.
.
$$g = gx + 2a_{n-2}$$
$$g = gx + a_{n-1}$$

which can be programmed as

```
g := n * a[0];
FOR i := 1 TO n-1 DO
   g := g * x + (n - i) * a[i];
```

These simple programming steps form the basis for two functions. As well as the value of x these also require the order of the polynomial n and the coefficients, in a, to be parameters. It should be noted that the function name to which the value is assigned must not be used in the summation process, otherwise the function will be called from itself. This will usually produce an error since the parameters will not match. Assuming that the array a has been declared to be of type coefficients, defined by the user, the functions will need to be something like the following:

```
FUNCTION f(x : REAL;
           n : INTEGER;
           a : coefficients):REAL;
VAR i : INTEGER;
    sum : REAL;
BEGIN
  sum := a[0];
  FOR i := 1 TO n DO
    sum := sum * x + a[i];
  f := sum;
END;

FUNCTION fdash(x : REAL;
              n : INTEGER;
              a : coefficients) : REAL;
VAR i : INTEGER; sum : REAL;
BEGIN
  sum := n * a[0];
  FOR i := 1 TO n-1 DO
    sum := sum * x + (n-i) * a[i];
  fdash := sum;
END;
```

These can then be called at the appropriate time in the Newton–Raphson procedures that we described in chapter 8.

Using this method, we can find a single real root to the polynomial $f(x)$. If a real root does not exist, then convergence will not occur. In this case we can accept that if after, say, 100 iterations, convergence has not occurred, then imaginary roots must be a possibility although this is not a certainty.

The next stage is to consider the problem of determining all the real roots of the polynomial.

Consider the derivative

$$f'(x) = na_0 x^{n-1} + (n-1)a_1 x^{n-2} + \ldots + a_{n-1}$$

$$= a_0 x^{n-1} + (a_0 x + a_1)x^{n-2} + ((a_0 x + a_1)x + a_2)x^{n-3} +$$

$$\ldots + ((\ldots((a_0 x + a_1)x + a_2)x + \ldots)\ldots + a_{n-2})x + a_{n-1}$$

$$= b_0 x^{n-1} + b_1 x^{n-2} + b_2 x^{n-3} + \ldots + b_{n-1}$$

$$\text{where} \quad
\left.
\begin{aligned}
b_0 &= a_0 \\
b_1 &= a_0 x + a_1 = b_0 x + a_1 \\
b_2 &= (a_0 x + a_1)x + a_2 = b_1 x + a_2 \\
&\quad\vdots \\
b_{n-1} &= b_{n-2}x + a_{n-1}
\end{aligned}
\right\} \tag{11.2}$$

Hence $b_0 b_1 b_2 \ldots b_{n-1}$ are the values at the successive stages of the polynomial evaluation given in equations (11.1) and

$$b_n = b_{n-1}x + a_n = f(x)$$

Suppose we have a set of values $b_0, b_1 \ldots b_{n-1}, b_n$ evaluated at an arbitrary value y using equations (11.2) and let

$$g(x) = b_0 x^{n-1} + \ldots + b_{n-1}$$

then

$$(x-y)\,g(x) = b_0 x^n + (b_1 - b_0 y)x^{n-1} + (b_2 - b_1 y)x^{n-2} + \ldots$$

$$\ldots (b_{n-1} - b_{n-2}y)x - b_{n-1}y$$

$$= a_0 x^n + a_1 x^{n-1} + a_2 x^{n-2} + \ldots + a_{n-1}x + a_n - b_n$$

We note that $b_n = f(y)$. Hence

$$(x-y)\,g(x) = f(x) - f(y)$$

Now if y is a root of the polynomial, $f(y) = b_n = 0$ approximately and

$$f(x) = (x-y)\,g(x)$$

gives a factorisation of $f(x)$ which removes the known root y. Hence, any further roots of $f(x)$ must also be roots of $g(x)$, where $g(x)$ is a polynomial of order $n-1$ whose coefficients are determined by equations (11.2) evaluated at the known root of $f(x)$.

Thus, if we save the values of the successive stages of the calculation of $f(x)$ in an array b, these can be used to

(a) evaluate the derivative used in determining the current root, and
(b) evaluate the function for determining the next root.

Thus, once we have determined a real root we reduce the order of the polynomial by 1 and transfer the values in b to a. We can then proceed to search for the next real root.

11.5.1 Construction of the program

We will need two arrays a and b with subscripts ranging from 0 to n. These are going to be used as parameters in procedures and functions, and hence need a related type declaration. For example

```
CONST maxorder = 5;
TYPE coefficients = ARRAY[0..maxorder] OF REAL;
VAR a,b : coefficients;
```

Note that *maxorder* specifies the maximum order of polynomial that the program can handle. This can be changed to suit the particular problem.

The steps in the main program are:

1. Read in the order of the polynomial.
2. Read in the coefficients of the polynomial.
3. Enter an initial guess at the first root.
4. Determine all the real roots.

This last step can be broken down further as follows:

4a. Determine a root of the polynomial using the Newton–Raphson method.
4b. IF a real root is found THEN
        display the root
        reduce the order of the polynomial
        assign the revised coefficients in b to a
     ELSE
        display a message that no further real roots can be found
     UNTIL either a real root was not found or the order becomes 1.
4c. IF order is 1 THEN
        determine the last root and display it.

We only need a guess at the first root if we use the previous root as an initial guess for the next.

To calculate the value of the polynomial we are going to need a procedure since, as well as the value of the polynomial, we will also require intermediate values for each value of x. These values will be stored in array b with the value of the polynomial being assigned to $b[n]$, hence b will be variable parameter of type *coefficients*. We will also need three value parameters: to receive the coefficients in a, the value x at which the polynomial is to be evaluated and n the order of the polynomial. The procedure is therefore:

```
PROCEDURE polyval(VAR b : coefficients;
                  a : coefficients;
                  x : REAL; n : INTEGER);
VAR i : INTEGER;
BEGIN
   b[0] := a[0];
   FOR i := 1 TO n DO
      b[i] := b[i-1] * x + a[i];
END;
```

The derivative of the polynomial can be calculated using a function, since we only require the single value. It will require three parameters: the vector b of type *coefficients* which receives the coefficient resulting from the procedure *polyval*, x a real to receive the value at which the evaluation is to be made, and n an integer to receive the order of the derivative. The form of the function is:

```
FUNCTION fdash(b : coefficients; x : REAL;
                           n : INTEGER) : REAL;
VAR i : INTEGER; sum : REAL;
BEGIN
   sum := b[0];
   FOR i := 1 TO n DO
     sum := sum * x + b[i];
   fdash := sum;
END;
```

The Newton-Raphson procedure which we developed in chapter 8 will only need minor modifications. Extra parameters will be required to handle the coefficients and the order of the polynomial. A variable parameter of type *coefficients* will be required for the revised set of coefficients, and value parameters — a of type *coefficients* and n of type INTEGER — will be needed to receive the original coefficients and the order of the polynomial. The heading of the procedure therefore needs to be revised to the following:

```
PROCEDURE newton(VAR currentvalue : REAL;
                 VAR b : coefficients;
                 a : coefficients;
                 VAR iterations : INTEGER;
                 nord : INTEGER;
                 VAR conv : BOOLEAN);
```

We will also need to replace the line

```
current := current - f(current) / fdash(current);
```

by two lines

```
polyval(b,a,currentvalue,nord);
currentvalue := currentvalue -
             b[nord] / fdash(b,currentvalue,nord-1);
```

If we put all this together we get the following program.

```
PROGRAM polyroots(INPUT,OUTPUT);
CONST maxord = 5;
TYPE coefficients = ARRAY[0..maxord] OF REAL;
VAR a,b : coefficients;
    inest : REAL;
    i, n, iterations : INTEGER;
    convergence : BOOLEAN;

PROCEDURE polyval(VAR b : coefficients;
                      a : coefficients;
                      x : REAL; n : INTEGER);
VAR i : INTEGER;
BEGIN
  b[0] := a[0];
  FOR i := 1 TO n DO
    b[i] := b[i-1] * x + a[i];
END;

FUNCTION fdash(b : coefficients; x : REAL;
                      n : INTEGER) : REAL;
VAR i : INTEGER; sum : REAL;
BEGIN
  sum := b[0];
  FOR i := 1 TO n DO
    sum := sum * x + b[i];
  fdash := sum;
END;

PROCEDURE newton(VAR currentvalue : REAL;
                 VAR b : coefficients;a : coefficients;
                 VAR iterations : INTEGER;
                 nord : INTEGER;
                 VAR conv : BOOLEAN);
CONST maxit = 100; epsilon = 0.0000001;
VAR oldvalue : REAL;
```

```
BEGIN
  iterations := 0;
  REPEAT
    oldvalue := currentvalue;
    polyval(b,a,currentvalue,nord);
    currentvalue := currentvalue -
                    b[nord]/fdash(b,currentvalue,nord-1);
    iterations := iterations + 1;
  UNTIL  (iterations = maxit) OR
                (ABS(currentvalue/oldvalue-1)<epsilon);
  conv := NOT(iterations = maxit);
END;

BEGIN

  WRITELN('what is the order of the polynomial');
  READ(n);
  WRITELN('what are the values of the coefficients');
  FOR i := 0 TO n DO
    READ(a[i]);
  WRITELN('enter an initial guess at the first root');
  READ(inest);

  (* step 4 determination of the REAL roots *)
  REPEAT
    (* step 4a *)
    newton(inest,b,a,iterations,n,convergence);
    (* step 4b *)
    IF convergence THEN
        BEGIN
            WRITELN('next real root is ',inest:12:4);
            n := n - 1;
            FOR i := 1 TO n DO
                a[i] := b[i];
        END
    ELSE
        WRITELN(' no real root found');
  UNTIL (n = 1) or NOT(convergence);

(* step 4c *)
  IF n = 1 THEN
      WRITELN('next real root is ',a[0]/a[1]:12:4);
END.
```

11.6 Exercises

11.1. Write a procedure to calculate the scalar product of two vectors.

11.2. Write a procedure to calculate the vector product of two vectors.

11.3. The trace of a matrix is defined to be the sum of its diagonal terms. Write a function to calculate the trace of an $n*n$ matrix.

11.4. Write a procedure to calculate the transpose of a matrix.

11.5. Write a procedure to calculate $A'A$ — that is, to pre-multiply a matrix A by its transpose.

11.6. Write a program which reads in a sequence of observations and then displays the lengths of the maximum ascending and descending runs in the data together with the positions of each peak and trough. (A peak is such that both its neighbours are less than it, while a trough has both of its neighbours greater than it.)

Modify the program to determine the average length between peaks, the average length of descending run and the average length of ascending run.

11.7. Write a program to sort an array of values into ascending order using the following algorithm (called exchange sort).

> starting with the first term in the array
> **REPEAT**
>> search through the part of the array not yet ordered
>>> and find the smallest term
>> swap this smallest term with first in this part of the array
>> move on to the next term in the array
> **UNTIL** all the array is ordered

11.8. A second sorting algorithm is the Bubble Sort method. In this we pass through the loop taking each pair of ordered subscripts in turn and swap them around if the values are out of order. If we perform this operation $n-1$ times in an array containing n values, then the array will be sorted. Thus the fundamental structure of the algorithm is

> **FOR** i := 1 **TO** n−1 **DO**
>> starting with the first pair
>> take each pair in turn and
>>> **IF** the values are in the wrong order **THEN** swap them

Write a program to sort an array using the Bubble Sort method.

Once an array is sorted, all subsequent passes through the array will be redundant. (The array is sorted and no more swaps will be made.) Modify your program to avoid redundant passes through the array.

Compare the speed of both versions of your program with that of the previous problem for a range of problems involving different numbers of terms.

12 Linear Algebra
—Simultaneous Equations

12.1 Introduction

A common problem in mathematics and its applications is that of solving a set of n simultaneous linear equations in n unknowns. That is, we want to solve

$$a_{11}x_1 + a_{12}x_2 + \ldots + a_{1n}x_n = b_1$$
$$a_{21}x_1 + a_{22}x_2 + \ldots + a_{2n}x_n = b_2$$
$$\bullet \qquad \bullet \qquad \qquad \bullet$$
$$\bullet \qquad \bullet \qquad \qquad \bullet$$
$$\bullet \qquad \bullet \qquad \qquad \bullet$$
$$a_{n1}x_1 + a_{n2}x_2 + \ldots + a_{nn}x_n = b_n$$

for $x_1, x_2 \ldots x_n$.

We can represent these in matrix form as $Ax = b$ where A is a $n*n$ matrix of coefficients a_{ij}, x is an $n*1$ vector of the n unknowns, and b is the $n*1$ vector containing the right-hand sides of the equations.

If the inverse of A exists, we can represent the solution algebraically as $x = A^{-1}b$. However, calculating the inverse of A is not the most efficient or most accurate way of solving this problem numerically on a computer. In this chapter we consider three methods for solving this problem, starting with the 'Gaussian elimination' method which solves it in the way we were first taught to do it by hand.

12.2 Gaussian Elimination

Consider the two variable problem

$$a_{11}x_1 + a_{12}x_2 = b_1 \qquad (1)$$
$$a_{21}x_1 + a_{22}x_2 = b_2 \qquad (2)$$

We solve this algebraically by using equation (1) to eliminate x_1 from equation (2). That is, we write

$$x_1 = b_1/a_{11} - x_2 a_{12}/a_{11}$$

and substituting in equation (2) gives us

$$x_2(a_{22} - a_{21}a_{12}/a_{11}) = b_2 - b_1 a_{12}/a_{11}$$

which we can solve for x_2 and hence obtain x_1 by substituting in (1).

In terms of the matrices defined above we start with

$$\begin{bmatrix} a_{11} & a_{12} \\ a_{21} & a_{22} \end{bmatrix} \begin{bmatrix} x_1 \\ x_2 \end{bmatrix} = \begin{bmatrix} b_1 \\ b_2 \end{bmatrix}$$

and the first step of our solution can be represented by the equations

$$\begin{bmatrix} a_{11} & a_{12} \\ 0 & a_{22} - a_{12}a_{21}/a_{11} \end{bmatrix} \begin{bmatrix} x_1 \\ x_2 \end{bmatrix} = \begin{bmatrix} b_1 \\ b_2 - b_1 a_{21}/a_{11} \end{bmatrix}$$

which we can write as

$$\begin{bmatrix} a_{11} & a_{12} \\ 0 & a_{22}* \end{bmatrix} \begin{bmatrix} x_1 \\ x_2 \end{bmatrix} = \begin{bmatrix} b_1 \\ b_2* \end{bmatrix}$$

From these we can obtain x_2 and x_1 by back-substitution.

We can overwrite the values in the matrix A and vector b in this way since we are not going to require the values again in the context of solving this problem.

The case of two equations in two unknowns illustrates the basic step. We next consider the three equation case.

The basic matrices are

$$\begin{bmatrix} a_{11} & a_{12} & a_{13} \\ a_{21} & a_{22} & a_{23} \\ a_{31} & a_{32} & a_{33} \end{bmatrix} \begin{bmatrix} b_1 \\ b_2 \\ b_3 \end{bmatrix}$$

We begin by eliminating variable 1 from equations (2) and (3) by subtracting from the terms in the 2nd and 3rd columns, and the right-hand side of each row, the term in the first column of that row multiplied by the value in the appropriate column in the first row divided by a_{11}. The values in the first column below the first row are replaced by zeros. Thus rows 2 and 3 become

$$\begin{bmatrix} 0 & a_{22} - a_{21}a_{12}/a_{11} & a_{23} - a_{21}a_{13}/a_{11} \\ 0 & a_{32} - a_{31}a_{12}/a_{11} & a_{33} - a_{31}a_{13}/a_{11} \end{bmatrix} \begin{bmatrix} b_2 - a_{21}b_1/a_{11} \\ b_3 - a_{31}b_1/a_{11} \end{bmatrix}$$

Thus the original values have been replaced by

$$\begin{bmatrix} a_{11} & a_{12} & a_{13} \\ 0 & a_{22}* & a_{23}* \\ 0 & a_{32}* & a_{33}* \end{bmatrix} \begin{bmatrix} b_1 \\ b_2* \\ b_3* \end{bmatrix}$$

The process of eliminating variable 1 from equations (2) and (3) has reduced these to two equations in two unknowns which can be solved as described earlier.

Finally the complete solution can be found by back-substitution.

We can generalise this to the general problem of n equations in n unknowns. Thus, if we have a $n*n$ system to solve, we eliminate x_1 from equations 2 to n by writing

> FOR $i:=2$ to n
> set multiplier $:= a_{i1}/a_{11}$
> FOR $j:=2$ to n
> $a_{ij}* = a_{ij} -$ multiplier $* a_{1j}$
> $b_i* = b_i -$ multiplier $* b_1$

The process is then repeated with the resulting $n-1$ equations until the last equation is in terms of x_n alone, which we can then solve and obtain the others by back-substitution.

It should be noted that the operations on the right-hand side are identical to those on the terms on the left-hand side of the equation. Hence, by storing the right-hand side as the $(n+1)$th column of the matrix A we do not have to bother about an extra vector.

We call the diagonal terms that we use as divisors, *pivotal terms* or *pivots*. The row and column containing the pivot are consequently called the *pivotal row* and the *pivotal column*.

This suggests the following as a basis for an algorithm:

> FOR rows i := 1 to n in turn
> eliminate the ith variable from rows i+1 to n by
> 1. determining the multiplier for that row then
> 2. subtracting the value in the appropriate column in the pivotal row multiplied by the multiplier from each value in columns i+1 to n

Clearly, the first step will cause problems if the pivotal value is zero. Therefore, before the first step we need to find the first row from the current pivotal row onwards which has a non-zero value in the pivotal column, and switch this row with the current pivotal row before proceeding.

Thus, before step 1 we need the following:

> IF pivotal value a$_{ii}$ is zero THEN
> search for the next row with a non-zero value in the ith column
> and swap these two rows

If all the values in the pivotal column are zero, then we cannot proceed. The matrix A is singular. In this case the equations either have no solution, or their solution is not unique.

The swapping process can be handled by a simple procedure for swapping two values. This will have two variable parameters to receive the values and return them swapped round. It will also need a local variable in which to store temporarily one value while the other is moved. The procedure will therefore be:

```
PROCEDURE swap(VAR first, second : REAL);
VAR temp : REAL;
BEGIN
   temp := first;
   first := second;
   second := temp;
END;
```

Returning to the main part of the algorithm, the inclusion of the test for singularity means that once we have established that our equations have no unique solution then there is no point in proceeding. Thus, rather than use a FOR loop it is better to use a REPEAT loop with two stopping conditions. That is, we want to stop either when we have reached the last equation or when we have identified the fact that our equations have no unique solution.

The revised algorithm for solving the problem is:

> set the singularity indicator to FALSE
> set the row counter i to 1
> REPEAT
> > 1. IF the pivotal value, a_{ii} is zero THEN
> > > 2. search for the next row with a non-zero value in the ith column
> > > 3. IF a non-zero value is found THEN
> > > > swap that row with the ith row
> > > > ELSE
> > > > > set the singularity indicator to TRUE
> > 4. IF the singularity indicator is not true THEN
> > > 5. FOR rows $k := i + 1$ to n
> > > > calculate the multiplier a_{ki}/a_{ii}
> > > > FOR cols $j := i + 1$ to $n + 1$
> > > > > set $a_{kj} = a_{kj} -$ multiplier$*a_{ij}$
> > > 6. increase the row number, that is $i := i + 1$
> UNTIL we have identified the singular case or reached the nth row

The last row is slightly different in that we only have to check for a non-zero coefficient a_{nn} and can then start the back-substitution. We therefore proceed as follows

> > 7. If a_{nn} is zero THEN set the singularity indicator to TRUE
> > > ELSE start the back-substitution
> > > > that is set $x_n = a_{n,n+1}/a_{nn}$
> > 8. If the singularity index is FALSE THEN
> > > > carry out the rest of the back-substitution.

The back-substitution is as follows:

> FOR $i := n-1$ DOWNTO 1
> > 1. set x_i equal to the right-hand side, divided by its pivotal value. That is $a_{i,n+1}/a_{ii}$

2. FOR j := i + 1 to n
      subtract $a_{ij}*x_j/a_{ii}$ from x_i

This leads us to the following code:

```
i := 1;
singular := FALSE;
REPEAT
  IF ABS(a[i,i]) < 0.0001 THEN
     BEGIN
 (* Step 2  search for  the  next  row  with a non-zero
            value in the ith column *)
        k := i + 1;
        WHILE ABS(a[k,i]) < 0.0001 DO
          k := k + 1;
(* Step 3  if a non-zero value is found *)
        IF k <= n THEN
(* Swap that row with the ith *)
           FOR j:= i TO n+1 DO
               swap(a[i,j],a[k,j])
        ELSE
(* set the singularity indicator*)
           singular := TRUE;
     END;

  IF NOT(singular) THEN
(*  Step  5 adjust values  in  the  remaining  rows  by
            subtracting from each term the value in the
            pivotal  column  times  the  value  in  the
            pivotal row *)
     FOR k := i + 1 TO n DO
        BEGIN
          multiplier := a[k,i] / a[i,i];
          FOR j := i + 1 TO n + 1 DO
          a[k,j] := a[k,j] - multiplier * a[i,j];
        END;

  i := i+1;
UNTIL singular OR (i = n);

(* Step 7 check the last equation and if it is not zero
          start the back substitution *)
  IF ABS(a[n,n]) < 0.0001 THEN singular := TRUE
  ELSE   x[n] := a[n,n+1] / a[n,n];

(*  Step  8 if a  solution  exists  complete  the  back
            substitution *)
  IF NOT(singular) THEN
     FOR i := n - 1 DOWNTO 1 DO
        BEGIN
          x[i] := a[i,n+1] / a[i,i];
          FOR j := i + 1 TO n DO
             x[i] := x[i] - a[i,j] * x[j] / a[i,i];
        END;
```

12.2.1 Numerical aspects

Consider the problem

$$\begin{bmatrix} 1 & 1/2 & 1/3 \\ 1/2 & 1/3 & 1/4 \\ 1/3 & 1/4 & 1/5 \end{bmatrix} \begin{bmatrix} x_1 \\ x_2 \\ x_3 \end{bmatrix} = \begin{bmatrix} 1 \\ 7/12 \\ 13/30 \end{bmatrix}$$

which, if we work in fractions, has the exact solution $1, -2, 3$.

However, on the computer we have to work in terms of real numbers which are truncated to a number of significant figures. For example, if we work the above problem to 5 significant figures the problem we solve is

$$\begin{bmatrix} 1.00000 & 0.50000 & 0.33333 \\ 0.50000 & 0.33333 & 0.25000 \\ 0.33333 & 0.25000 & 0.20000 \end{bmatrix} \begin{bmatrix} x_1 \\ x_2 \\ x_3 \end{bmatrix} = \begin{bmatrix} 1.00000 \\ 0.58333 \\ 0.43333 \end{bmatrix}$$

which has solution $1.0045 \quad -2.0229 \quad 3.0211$.

If we had worked to 4 significant figures the answer would be $0.9971 \quad -1.982$ 2.952. The accuracy that can be achieved depends on the particular computer being used and hence techniques need to be developed which can help overcome this problem.

Consider a second problem

$$\begin{bmatrix} 0.00031 & 1 \\ 1 & 1 \end{bmatrix} \begin{bmatrix} x_1 \\ x_2 \end{bmatrix} = \begin{bmatrix} -3 \\ -7 \end{bmatrix}$$

If we solve the equations as they stand using 0.00031 as the pivot, we get, working to 4 significant figures

$$\begin{bmatrix} 1 & 3226 \\ 0 & -3225 \end{bmatrix} \begin{bmatrix} x_1 \\ x_2 \end{bmatrix} = \begin{bmatrix} -9677 \\ 9670 \end{bmatrix}$$

leading to $x_1 = -5.452$ and $x_2 = -2.998$, which clearly has a large error in the context of the second equation. However, if we rearranged the equations and used 1 as our pivot then the solution is $-4.001 \quad -2.998$, which clearly gives a more accurate result.

This suggests an immediate improvement to our Gaussian elimination algorithm, namely that we should select as our pivot the largest in absolute value of those available.

Thus, for each row in turn we should search over the remaining rows for the largest absolute value in the pivotal column and use that as the pivot by swapping the rows. If the largest absolute value is zero then all the possible pivots are zero and there is no unique solution, otherwise we can proceed. If we now build this into our algorithm, we have the following steps:

set singularity indicator to **FALSE**
set row number i to 1

REPEAT

      search for the largest absolute value in the pivotal column over
           rows i to n and identify the row, k say

      IF the largest is not zero THEN

          1. IF $k > i$ THEN swap rows i and k

          2. eliminate variable i from rows i+1 to n as before

      ELSE set singular := TRUE

      move on to the next elimination by increasing the line number

UNTIL we have identified the singular case or reached the last line

Check the last row and if a_{nn} is non-zero start the back-substitution

IF there exists a solution perform a back-substitution to obtain the result.

This leads us to the following code:

```
singular := FALSE;
i := 1;
REPEAT
   (*  search  for the largest value in the  rest of the
       pivotal column *)
   k := i;
   maximum := ABS(a[i,i]);
   FOR j := i TO n DO
     IF ABS(a[j,i]) > maximum THEN
        BEGIN
           maximum := ABS(a[j,i]);
           k := j;
        END;

   IF maximum > eps THEN
     BEGIN
(*  if  a  non zero  maximum  is  found  swap  rows  if
     necessary *)
       IF k > i THEN
          FOR j := i TO n + 1 DO
             swap(a[i,j],a[k,j]);
   (* adjust the corresponding part of the A matrix *)
       FOR k:=i+1 TO n DO
          BEGIN
             multiplier := a[k,i] / a[i,i];
             FOR j := i + 1 TO n + 1 DO
                a[k,j] := a[k,j] - multiplier * a[i,j];
          END;
     END

   ELSE
   (* else set the singularity indicator true *)
     singular:=TRUE;

   i:=i+1;
UNTIL singular OR (i=n);
```

The remainder of the program which handles the last row and the back-substitution is the same as before.

12.2.2   Constructing a procedure

Once the structure of the code to solve the problem has been formulated, we can construct a procedure which can then be used in any context where the solution of simultaneous equations is required.

The procedure will need four parameters

x	a variable parameter of $n*1$ vector type in which the solution is placed
a	a value parameter of $n*(n+1)$ matrix type which on entry to the procedure will contain the A matrix and the right-hand sides
n	a value parameter of integer type to receive the number of unknowns
singular	a variable parameter of Boolean type which will have the value FALSE if a solution exists and TRUE if there is not a unique solution

All the other identifiers used in the procedure will be declared as local identifiers.

Thus the last version of our program can be placed in a procedure headed:

```
PROCEDURE gauss(VAR x : vector; a : matrix;
                n : INTEGER;
                VAR singular : BOOLEAN);
CONST eps = 1E-5;
VAR i,j,k : INTEGER; max,multiplier : REAL;
```

12.3   Iterative Methods

The number of operations performed in using Gaussian elimination can be shown to be approximately $2n^3/3$ for an $n \times n$ system, clearly making it a lengthy process. Furthermore, since successive stages of the calculation use previously calculated values, rounding errors may accumulate. We therefore introduce an alternative method of solution which can achieve greater accuracy and in some cases use fewer operations.

The method uses the idea of a sequential mapping which can be used to generate a sequence of values starting from a given initial value. Applying this idea to the problem of finding a solution of a non-linear equation we had an iterative process which under certain conditions converged. The examples we considered in chapters 4 and 8 were all for a single-variable problem, but the methods can be easily extended to situations involving more than one variable and more than one equation.

One method of finding a solution to the equation $f(x) = 0$ was to rearrange the equation in the form $x = g(x)$ and this could be used to generate a sequence of values, according to the rule $x_{k+1} = g(x_k)$, which under certain conditions converged. Suppose we now apply this technique to our system of n linear equations given by

$$a_{11}x_1 + a_{12}x_2 + \ldots + a_{1n}x_n = b_1$$
$$a_{21}x_1 + a_{22}x_2 + \ldots + a_{2n}x_n = b_2$$

$$\cdot \qquad \cdot \qquad \qquad \cdot$$
$$\cdot \qquad \cdot \qquad \qquad \cdot$$
$$\cdot \qquad \cdot \qquad \qquad \cdot$$

$$a_{n1}x_1 + a_{n2}x_2 + \ldots + a_{nn}x_n = b_n$$

Dividing each equation by the diagonal term a_{ii}, we can rewrite them as

$$x_1 = -\frac{a_{12}x_2}{a_{11}} - \ldots - \frac{a_{1n}x_n}{a_{11}} + \frac{b_1}{a_{11}}$$

$$x_2 = -\frac{a_{21}x_1}{a_{22}} - \ldots - \frac{a_{2n}x_n}{a_{22}} + \frac{b_2}{a_{22}}$$

$$\vdots \qquad \vdots \qquad \qquad \vdots \qquad \qquad \vdots$$

$$x_n = -\frac{a_{n1}x_1}{a_{nn}} - \ldots - \frac{a_{n-1,n-1}x_{n-1}}{a_{nn}} + \frac{b_n}{a_{nn}}$$

If we write this as $x = -A^*x + b^*$ where

$$A^* = \begin{bmatrix} 0 & a_{12}/a_{11} & \ldots & a_{1n}/a_{11} \\ a_{21}/a_{22} & 0 & \ldots & a_{2n}/a_{22} \\ \cdot & \cdot & & \cdot \\ \cdot & \cdot & & \cdot \\ \cdot & \cdot & & \cdot \\ a_{n1}/a_{nn} & a_{n2}/a_{nn} & \ldots & 0 \end{bmatrix} \quad b^* = \begin{bmatrix} b_1/a_{11} \\ b_2/a_{22} \\ \cdot \\ \cdot \\ \cdot \\ b_n/a_{nn} \end{bmatrix}$$

we can use this as a sequential mapping of the form

$$x_{k+1} = -A^*x_k + b^*$$

to generate a sequence of estimates of the vector x which under certain conditions may converge to the solution of the system of equations.

We can either

(1) start from a vector x_k and generate a complete new vector x_{k+1} using the n equations (this is known as the Jacobi method), or

(2) use each equation in turn to update that term in the vector x (this is known as the Gauss–Siedel method).

12.3.1   The Jacobi method

The Jacobi method uses the sequential mapping

$$x(k+1) = -A^*x(k) + b^*$$

to generate a new estimate of the vector x from the current estimate. Suppose we begin by considering under what conditions we can expect convergence.

If x is the true answer and $x(k)$ and $x(k+1)$ are succeeding values in the sequence, then

$$x(k) = x + E(k)$$

and

$$x(k+1) = x + E(k+1)$$

where $E(k)$ and $E(k+1)$ are the respective errors. If all the absolute errors in $E(k+1)$ are less than the maximum absolute error in $E(k)$, then the sequence of values can be said to be converging. Now

$$x(k+1) = x + E(k+1)$$

$$= -A*(x + E(k)) + b*$$

$$= -A*x + b* - A*E(k)$$

$$= x - A*E(k)$$

Hence $E(k+1) = -A*E(k)$

That is $\quad |E_i(k+1)| = | \sum_j a_{ij}^* E_j(k)|$

Therefore $\quad |E_i(k+1)| \leqslant | \sum_j a_{ij}^*| \, |E_j(k)|$

$$\leqslant \max |E_j(k)| \sum_j |a_{ij}^*|$$

Now $\quad \sum_j a_{ij}^* = \sum_{j \neq i} a_{ij}/a_{ii}$ since a_{ii}^* is zero

hence if $\quad \sum |a_{ij}| \leqslant |a_{ii}|$ then $\sum |a_{ij}^*| \leqslant 1$

and $\quad |E_i(k+1)| \leqslant \max |E_j(k)|$

Thus, if the absolute value of the diagonal terms in the matrix A are all greater than or equal to the sum of the absolute values of the terms in the corresponding row, then the iteration converges. A matrix with this property is said to be diagonally dominant.

Thus, before applying the sequential mapping to perform the iteration we need to check whether the matrix A is diagonally dominant. If it has this property then the iteration will converge to the solution of the equations. However, if A is not diagonally dominant it does not necessarily mean that convergence will not occur, but no guarantee can be given either way. Thus, if A is not diagonally dominant we usually do not proceed with this method.

12.3.2 The Gauss–Siedel method

The Gauss–Siedel method is a slightly improved version which can converge more quickly than the Jacobi method. The difference this time is that the equations are used in cyclic order with the vector x being revised after the use of each equation. Thus equation (1) is used to revise the estimate of x_1. Equation (2) is then used to revise x_2. This process is repeated for each equation in turn until x_n has been revised. We then return to equation (1), continuing until convergence has occurred.

Convergence is said to have occurred when the absolute changes in the x values on one sweep through the equations are all less than some pre-set level of accuracy.

We can illustrate this technique using the following pair of equations:

$x-y=0$ and $x+2y=2$

Suppose we arrange these as $x=y$ and $y=(2-x)/2$ and start from $x=3$ and $y=2$, then the iteration goes

		x	y
		3	2
$x=y$	$\Rightarrow$	2	
$y=(2-x)/2$	$\Rightarrow$		0
$x=y$	$\Rightarrow$	0	
$y=(2-x)/2$	$\Rightarrow$		1
$x=y$	$\Rightarrow$	1	
$y=(2-x)/2$	$\Rightarrow$		1/2
$x=y$	$\Rightarrow$	1/2	
$y=(2-x)/2$	$\Rightarrow$		3/4
$x=y$	$\Rightarrow$	3/4	
$y=(2-x)/2$	$\Rightarrow$		5/8
$x=y$	$\Rightarrow$	5/8	
$y=(2-x)/2$	$\Rightarrow$		11/16
		etc.	

until we reach the solution $x=2/3$, $y=2/3$. The iteration is illustrated in figure 12.1.

As with the Jacobi method, Gauss–Siedel only converges under certain conditions. It can be shown that if A is diagonally dominant the convergence is guaranteed, although again it is not a necessary condition.

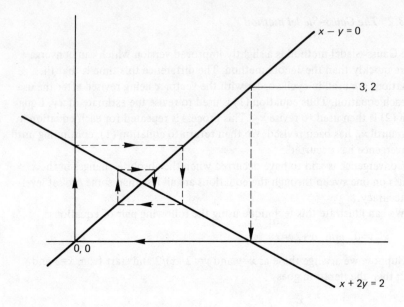

Figure 12.1

12.3.3 *Programming considerations*

The Jacobi method consists of the following steps

1. Check whether A is diagonally dominant.
2. IF A is diagonally dominant THEN
        (a)   replace A by A^* and b by b^*
        (b)   perform the iterations

Step 1. A is diagonally dominant if $|a_{ii}| \geqslant \Sigma |a_{ij}|$ for all i.
   We therefore need to check the terms

$$d = a_{ii} - \Sigma\, a_{ij} \text{ for } i = 1 \text{ to } n$$

until the first one becomes negative. If none becomes negative, then the matrix A is diagonally dominant.

   We can perform this most efficiently using two nested **REPEAT** loops as follows:

```
set i = 0
REPEAT
(* consider each row in turn until all rows have been checked or a
   negative value of d is reached *)
   set i = i+1
   set d = ABS(a[i,i])
   set j = 0
```

REPEAT
   (∗ subtract each non-diagonal term in turn until all have been
   included or a negative value of *d* is obtained ∗)
      set $j = j + 1$
      IF *i* is not equal to *j* THEN subtract $|a[i,j]|$ from *d*
      UNTIL $j = n$ or $d < 0$
   UNTIL $i = n$ or $d < 0$
   set diag = NOT$(d < 0)$ (∗ diag is a BOOLEAN variable which is TRUE
                           if A is diagonally dominant ∗)

Step 2a. Replace the matrix *A* by *A*∗ and the right-hand side *b* by *b*∗.
   This is achieved by two nested FOR loops as follows:

FOR each row *i* in turn
   set $d = a[i,i]$ the diagonal term
   set $b[i] = b[i]/d$
   set the diagonal term $a[i,i]$ = zero
   divide the remainder of the row by *d* the diagonal term
      That is   FOR *j*= 1 to *n*
                   IF *j* not equal to *i* THEN
                      replace $a[i,j]$ by $a[i,j]/d$

Step 2b. We perform the iterations by repeating the iterative step until convergence occurs. (Note that we only carry out this step if the matrix *A* is
   diagonally dominant and the convergence is guaranteed.) The steps are:

REPEAT
   store the current estimate of *x* in *y*
   calculate the new estimate in *x* by using
      $x = -A*y + b*$
   Check for convergence
UNTIL convergence has occurred.

The calculation of the revised estimate simply requires two nested FOR loops
and is shown in the full procedure given at the end of this section.

The check for convergence is carried out as follows:

Start by setting a BOOLEAN variable identifier convergence = TRUE. Check
each estimate in turn to see if it is within epsilon of the previous estimate until
either all have been checked or one has been found which is not. If the latter
case is found, set convergence = FALSE. This has the following structure:

set convergence = TRUE
set $i = 0$
REPEAT
   set $i = i + 1$
   IF the absolute difference between estimates
      that is ABS$(x(i)-y(i))$ is greater than epsilon THEN
         set convergence = FALSE

UNTIL $(i = n)$ OR NOT(convergence)

That is UNTIL all terms have been checked or the first one is
found that makes convergence FALSE

To complete the procedure we need to decide on the parameters. Two pieces
of information are required at the end of the iteration: the answer which will
be in the REAL *vector x* and the indication as to whether convergence is possible
which is in the BOOLEAN variable *diag*. These therefore will need to be variable
parameters. The procedure also has to receive information about the *matrix A*,
the right-hand side of which is in the *vector b* and the number of equations in
the INTEGER variable *n*. These three will be value parameters which has the
advantage that, although *A* and *b* are changed in the procedure, this will not
result in them being altered in the part of the program that calls the procedure.

If we now put all this together, we get the following procedure for performing the Jacobi method.

```
PROCEDURE jacobi(VAR x : vector;
                 a : matrix; b : vector;
                 n : INTEGER; VAR diag : BOOLEAN);
CONST epsilon = 0.00001;
VAR i,j : INTEGER; d : REAL; y : vector;
    convergence : BOOLEAN;

BEGIN
(*  step 1.  Check whether  the  matrix  is  diagonally
             dominant *)
  i := 0;
  REPEAT
    i := i + 1;
    d := ABS(a[i,i]);
    j := 0;

    REPEAT
      j := j + 1;
      IF i <> j THEN d := d - ABS(a[i,j]);
    UNTIL (j = n) OR (d < 0);

    diag := NOT(d < 0);
  UNTIL (i = n) OR NOT(diag);

  IF diag THEN
(* step 2a construct the matrix A* and the vector b* *)
      BEGIN
        FOR i := 1 TO n DO
          BEGIN
            d := a[i,i];
            a[i,i] := 0.0;
            b[i] := b[i] / d;
            FOR j := 1 TO n DO
              IF j <> i THEN a[i,j] := a[i,j]/d;
          END;
```

```
(* step 2b perform the iterations *)
    REPEAT
      FOR i := 1 TO n DO
        y[i] := x[i];
      FOR i := 1 TO n DO
        BEGIN
          x[i] := b[i];
          FOR j := 1 TO n DO
            IF i <> j THEN x[i] := x[i] -
                                   a[i,j] * y[j];
        END;
      convergence := TRUE;
      i := 0;
      REPEAT
        i := i + 1;
        IF ABS(x[i] - y[i]) > epsilon THEN
                  convergence := FALSE;
      UNTIL (i = n) OR NOT(convergence);
    UNTIL convergence;
  END;
END;
```

The Gauss–Siedel method can be constructed in a similar way, the only difference coming at the iteration stage. We only need to store a single value at each stage of the iteration and this is tested for convergence before moving on to the next term in the vector of unknowns. This process is repeated for each of the n terms in turn. The whole process is then repeated until overall convergence is obtained.

This is achieved using a structure such as the following:

> REPEAT
> > set convergence = TRUE
> > (∗ revise all the estimates in turn checking each for convergence ∗)
> > set row number $i = 0$
> > REPEAT
> > > set $i = i + 1$
> > > store $x[i]$ in y
> > > calculate the revised estimate for $x[i]$ using the vector x in
> > > > place of the vector y used in the Jacobi method.
> > >
> > > IF the absolute difference between the old estimate y and the
> > > > original estimate $x[i]$ is greater than epsilon
> > > > THEN set convergence = FALSE
> > >
> > UNTIL $i = n$
> UNTIL convergence

We therefore use a procedure virtually the same as before with step 2b replaced by

```
REPEAT
  convergence := TRUE;
  i := 0;
  REPEAT
    i := i + 1;
    y := x[i];
    x[i] := b[i];
    FOR j := 1 TO n DO
      IF i <> j THEN x[i] := x[i] - a[i,j] * x[j];
    IF ABS(x[i]-y) > epsilon THEN
            convergence := FALSE;
  UNTIL (i = n);
UNTIL convergence;
```

The identifier y in this case needs only to be defined as a **REAL** rather than a vector.

12.4 Exercises

12.1. Apply the methods described in this chapter to solving the following two problems:

(a) solve $\sqrt{2}x + \pi y = \sqrt{3}$
$\qquad ex + \sqrt{7}y = \sqrt{5}$

(b) solve $\Sigma \, (-1/j)^k x_k = 1$ $(k = 0,1\ldots n)$ for $n = 1,2,3$.

12.2. After the Gaussian elimination method has been carried out, the determinant of the matrix A is $(-1)^d \prod\limits_{j=1}^{a} a_{jj}^*$ where d is the number of changes of lines carried out in the process. Write a program to calculate the determinant of the matrix A.

12.3. Write a program to calculate the inverse of a matrix using the procedure to calculate its determinant from exercise **12.2**. Use it to calculate the inverse of the matrix

$$\begin{bmatrix} 1 & 1 & 1 & 1 \\ 1 & 2 & 4 & 8 \\ 1 & 3 & 9 & 27 \\ 1 & 4 & 16 & 64 \end{bmatrix}$$

12.4. Modify the Gaussian elimination program so that it can solve the system of equations with more than one right hand. That is, to solve the system of equations

$$A X = B$$

where A is an $n*n$ matrix, X is an $n*m$ matrix of unknowns and B is an $n*m$ matrix of right-hand sides. (If $m = 1$ we have the simultaneous equation problem we have already solved.)

If $n = m$ and we let $B = I$, the identity matrix, then the solution is

$$X = A^{-1}B = A^{-1}I = A^{-1}$$

giving us the inverse of the matrix A.

Write a program to solve the general system of equations and to calculate the inverse of the matrix A using this method.

13 Characters and User-defined Scalar Types

13.1 Characters

The only variables we have used so far have been numerical or Boolean quantities, with the exception of the use of strings of characters between apostrophes in write statements. These strings have been used for including text in our output to make it clearer to understand or to supply understandable prompts when our programs are expecting data to be entered at the terminal.

It is sometimes useful, though, to be able to manipulate text within a program. Applications of this are extensive and include word processing, lineprinter or VDU graphics, algebraic manipulation, and the construction and identification of keywords for selecting operations in menu-driven or word-driven packages.

Later, we will consider examples of lineprinter graphics but will not have space to cover the other potentially more interesting and useful applications.

In general, text is made up of strings of characters. We begin by introducing a variable type which can handle single characters. Characters may be assigned to identifiers of type CHAR. The values that may be assigned are elements of a finite and ordered set of characters. The actual set available and the ordering on them is computer dependent and the complete set can vary from one machine to another.

It is useful, though, if there is a minimal set of assumptions that can be made about any character set and hence the following ground rules exist in Pascal.

The character set must include:

1. the alphabetically ordered set of capital letters A. . .Z
2. the numerically ordered and contiguous set of decimal digits 0. . .9 and
3. the blank character.

In addition many compilers allow:

4. the alphabetically ordered set of small latin letters a. . .z

Note that while 1 and 4 require ordering in the natural way, they are not necessarily neighbours in the full ordered list of characters. However, the digits are expected to be neighbours in the ordered list.

The examples we have already met are constants of the type that can be represented between apostrophes. For example, '*', 'G', '3', 'a'.

Two standard functions for use with character variables are available which make use of this underlying ordering of the set. They are

ORD(c) which returns the ordinal number of the character currently
assigned to c

CHR(i) which returns the character corresponding to the ordinal number
currently assigned to the integer i

We should note here that for digits, ORD('3') for example does not return the value of 3. If we require to transfer information directly from characters to numbers then ORD('3')–ORD('0') will return the value 3.

Because characters can take elements from an ordered set we can compare them using the relational operators $<, >, \geqslant, \leqslant, \langle \rangle, =$ in a sensible way.

Thus, if c1 and c2 are two character identifiers then

$$c1 < c2 \text{ if and only if } ORD(c1) < ORD(c2)$$

Two other functions that are useful for character variables are **PRED** and SUCC. Their use is not restricted to character variables and can, for example, be used for incrementing integer variables used for counting.

PRED(c) which returns the previous character in the ordered list

SUCC(c) which returns the next character in the ordered set

Thus SUCC('b') is the character c while PRED('f') is the character e.

However, if the resulting value is outside the range of characters then an error will result.

Character variables can be assigned values.

1. In assignment statements. For example
   c := 'a' assigns the letter a to the variable c
2. Through the read statement. For example
   READ(first,second,third) will read the first three characters on
   the input medium (file or terminal) into the variables *first, second*
   and *third* (blanks are treated as characters and not used to separate
   strings as is the case for integer and real values)
3. Characters can also be assigned to constants. For example
   CONST blank = ' '; comma = ',';

A character variable can only be assigned a single letter, with the exception of those assigned in the constant statement.

Thus

CONST equation = 'a + b ∗ x';
.
.
.
WRITELN(equation);

is a valid use of characters and produces the output a + b * x, but c := 'a + b * x'
is invalid for an assignment statement.

Character variables require a single format number when being output, which
is interpreted in the same way as that used for integer variables. It allocates the
number of characters that will be displayed in the output, inserting blanks
before the characters currently assigned to the variable where necessary. For
example

> c := 'a';
> WRITELN(c:5) would produce an a preceded by four blanks
>
> WRITELN('a':5) would produce the same result

To handle strings of characters we need to use ARRAYs OF CHAR and, in
particular, a special type called PACKED ARRAY which we are not going to
describe in this text.

13.2   Scalar Types

We have now met three examples of scalar types of identifier that can be used
in Pascal. They are identifiers that can take values from an ordered, discrete,
finite set of values. The three we have met are INTEGER, BOOLEAN and
CHAR. In Pascal we are also allowed to declare our own scalar types. We can
do this by either restricting the range of an existing type or listing all the values
in the type. In both cases we are specifying the values that variables of the
declared type can be assigned.

13.2.1   Subrange types

One way in which we define our own types is by specifying a type as a subrange
of an existing type. Variables of this new type can only take values in the
specified range. For example

```
TYPE   bound = 1800..2500;
       smallletter = 'a'..'z';
```

declares a type *bound* which can only take integer values from 1800 to 2500
and a type *smallletter* that can only take lower-case latin letters.

One obvious application of this is in declaring bounds of arrays. Thus, for
example, the following definitions

```
CONST maxd = 1200;
TYPE bound = 1..maxd;
VAR x : ARRAY[bound] OF REAL;
    i : bound;
```

declare an array which can take subscripts in the range 1 to 1200 and an integer identifier that can only take values in the range of valid subscripts. Thus, if we attempt to assign a value to the subscript outside this range an error will occur and be indicated on the variable i rather than at the time it is used as a subscript.

Thus an advantage in declaring arrays in this way is that we can avoid the high and low bound checking errors which occur when the subscript value is out of the declared range for the array and obtain much more useful error messages.

The range of values that a variable can take can also be declared in the **VAR** statement. For example

VAR digits : 0. . .9;

declares a variable *digits* that can take values in the range 0. . .9. However, this method means that we cannot use the range, without restating it, in other variable definitions.

13.2.2 Enumerated types

Scalar types can also be declared in terms of the ordered list of values that they can take. We call these enumerated types and they have the advantage of representing quantities which can only take values within a declared set of values. It means that errors can be avoided, such as assigning wrong values to variables of the particular type.

Examples

```
TYPE sex : (female,male);
     weekday : (monday,tuesday,wednesday,thursday,
                                   friday,saturday,sunday);
     colour : (red,blue,green,yellow);

VAR person,student,employee : sex;
    hue : colour;
    day : weekday;
```

The order implied in the definition of this type of variable implies that enumerated types can be used as the control loop, the values passing over the set between the stated limits. For example

```
FOR day := monday TO friday DO
```

means that the statements in the loop are obeyed 5 times, once for each day in the working week.

They can also be used in conditions in IF constructions and REPEAT and WHILE loops. Thus, for example, the following are valid using the variables defined above:

```
IF student = male THEN ...

WHILE hue <> red DO ...

IF day > friday THEN ...
```

The order of the values corresponds to that used in the declared list. Values occurring later in the list are assumed to have a higher value than preceding values. Thus our final example of a condition is true if and only if *day* has the values *saturday* or *sunday*.

The main disadvantage with enumerated types is that values cannot be read into, or output directly from, them. For example, for the above examples we cannot use statements such as READ(student) or WRITE(day).

The output of values requires, for example, use of the case statement. Thus, to output the values of *day* requires a construction such as

```
CASE day OF
   monday :              WRITE('Monday');
   tuesday :             WRITE('Tuesday');
   wednesday :           WRITE('Wednesday');
   thursday :            WRITE('Thursday');
   friday :              WRITE('Friday');
   saturday,sunday :     WRITE('weekend');
END;
```

The entry of values requires a similar construction in order to convert numerical or character values that can be read into values of the defined type. Thus, if *daycode* has a declared subrange of values 1...7 then we could use the following construction to assign values to *day*.

```
READ(daycode);
CASE daycode OF
   1 : day := monday;
   2 : day := tuesday;
   . . . .
   7 : day := sunday;
END;
```

13.3 Character Graphics

As an example of the use of characters, we consider the problem of producing a lineprinter plot or screen plot of a function $f(x)$. The accuracy of such a plot is limited by the number of available plotting positions. The screen on a VDU, for example, has 80 character positions across the screen and 23 down the screen while a standard lineprinter has 132 characters across the page and 60 down the page. In both cases a greater depth can be obtained by allowing the top of the graph to move off the top of the screen as it scrolls up or the graph to run over more than one page.

Thus, any graphics procedure that is to be generally useful will require us to specify the *screenwidth* and *screendepth*, thus designating the grid into which we are going to draw our approximation to the curve. The graph will be an approximation since any point will be placed in the middle of the rectangle in the grid in which the *x,y* value lies.

The most flexible approach to drawing pictures in this way is to use an array of type character which represents the grid. The positions in the array are all set to the blank character and then only those which correspond to points on the function we are drawing are reset to another character which represents our graph.

If a graph of $y=f(x)$ is being drawn over an interval $[a,b]$, then we need to calculate the function at intervals of size $(b-a)$/screenwidth.

We shall represent each point on the curve by a '*', and the next step is to work out in which position to place these characters.

Obviously over the range of *x* values we are plotting the function $f(x)$ will have a maximum and minimum which will be the limits of the range of *y* values we will require. We could either supply this information through a read statement or write a procedure that calculates these two values which we shall call *minf* and *maxf*.

Once we know these two values then the *y*-axis is divided into intervals, where the length of the interval is $(maxf-minf)$/screendepth. The position of any point in the grid of screen points is determined by the interval on the *y*-axis in which the value of $f(x)$ lies.

Thus $j = \text{ROUND}((f(x)-minf)/\text{interval})$, where interval is $(maxf-minf)$/screendepth, is the interval number in which $f(x)$ lies.

The problem of drawing a character graph can therefore be broken down into the following steps:

1. Define the screensize
    that is, the number of positions available for characters.
2. Set the array that represents the screen to blanks.
3. Set up range and interval of the *x* values.
4. Determine the maximum and minimum values of the *y* variable.
5. Draw the axis.
6. Determine the positions in the screen array in which points on the curve lie and assign the appropriate character.
7. Display the graph.

We will write each of these as procedures called *screensize, blanks, xscale, minmax, axis, graph* and *displayscreen* respectively.

The screensize and the array in which the picture is to be stored will be global parameters called *screenwidth, screendepth* and *picture* respectively. The two dimensions of *picture* will run from 0 to *screenwidth* and 0 to *screendepth* respectively, and we will use 0,0 as the bottom left-hand corner of the screen, thus making our array represent the screen in the more natural way of visualising it. The variables *screenwidth* and *screendepth* are declared to be of subrange type, so that their values are restricted to a sensible range. Thus the maximum depth of picture is set to 60 while the maximum width is set to 132, corresponding to the standard lineprinter paper size.

```
TYPE depthbound = 0..59; widthbound = 0..131;
     screen = ARRAY[depthbound,widthbound] OF CHAR;
VAR  screenwidth : widthbound;
     screendepth : depthbound;
     picture : screen;
```

Procedure *screensize* requests the user to supply the screen dimensions.

```
PROCEDURE screensize;
BEGIN
    WRITELN('enter values for the depth',
                   ' and width of your picture');
    READLN(screendepth,screenwidth);
END;
```

Procedure *blanks* assigns a blank character to every location in the screen array that is going to be used.

```
PROCEDURE blanks;
VAR i : depthbound; j : widthbound;
BEGIN
    FOR i := 0 to screendepth DO
        FOR j := 0 to screenwidth DO
            picture[i,j] := ' ';
END;
```

Procedure *xscale* requests the boundaries to the x interval that is required and uses these to determine the interval, on the x-axis, between points that are going to be plotted.

```
PROCEDURE xscale(VAR lowbd, highbd, interval : REAL);
BEGIN
    WRITELN('enter the lower and',
                ' upper boundaries of the x interval);
    READLN(lowbd, highbd);
    interval := (highbd - lowbd) / screenwidth;
END;
```

Procedure *minmax* determines the minimum and maximum of the function f(*x*) over the range of *x* values.

```
PROCEDURE  minmax(VAR  max,min  :  REAL;
                  lowbd, interval : REAL; n : widthbound);
VAR x, y : REAL;
    i : widthbound;
BEGIN
    x  :=  lowbd;  (*  lower  bound  of the range of x
values *)
    min := f(x);
    max := f(x);
    FOR i := 1 to n DO
        BEGIN
            x := x + interval;
            y := f(x);
            if min > y THEN    min := y
                ELSE    IF max < y THEN max := y;
        END;
END;
```

Procedure *axis* inserts a '-' in the appropriate position for the *x*-axis and a ':' for the *y*-axis.

```
PROCEDURE axis(highbdx, lowbdx, intervalx, highbdy,
                       lowbdy,intervaly : REAL);
VAR i, centrex : widthbound; j, centrey : depthbound;
BEGIN
(* determine the row into which the x axis is placed *)
    IF (0 > lowbdy) AND (0 < highbdy) THEN
        centrey := ROUND((0 - lowbdy) / intervaly)
    ELSE
        centrey := 0;
(* determine the column into which the y axis is placed
*)
    IF (0 > lowbdx) AND (0 < highbdx) THEN
        centrex := ROUND((0 - lowbdx) / intervalx)
    ELSE
        centrex := 0;
    FOR j := 0 to screenwidth DO
        picture[centrey,j] := '-';
    FOR i := 0 to screendepth DO
        picture[i,centrex] := ':';
END;
```

Procedure *graph* stores a * in the locations of the screen array that correspond to points on the curve.

```
PROCEDURE graph(maxf, minf, lowbd, intervalx,
                                   intervaly : REAL);
VAR i : depthbound; j : widthbound;
    x : REAL;
BEGIN
  x := lowbd;
  FOR j := 0 TO screenwidth DO
    BEGIN
      i := ROUND((f(x) - minf) / intervaly);
      picture[i,j] := '*';
      x := x + intervalx;
    END;
END;
```

Procedure *displayscreen* outputs the array *screen* either to the screen or a file, depending on the response to a prompt which appears on the screen. It is basically the procedure *displaymat* used in chapter 11, except that the rows are printed out in reverse order.

```
PROCEDURE displayscreen;
VAR i : depthbound; j : widthbound;
    ans : CHAR;
    acceptableanswer : BOOLEAN;
BEGIN
  WRITELN('Do you want the display on',
                   ' the screen or in a file');
  REPEAT
    READLN(ans);
    acceptableanswer := (ans = 'S') OR (ans = 's')
                       OR (ans = 'F') OR (ans = 'f');
    IF NOT acceptableanswer THEN
         WRITELN('answer S or F');
  UNTIL acceptableanswer;
  FOR i := screendepth DOWNTO 0 DO
    BEGIN
      FOR j := 0 to screenwidth DO
        IF (ans = 's') OR (ans = 'S') THEN
          WRITE(picture[i,j])
        ELSE
          WRITE(result,picture[i,j]);
      IF (ans = 's') OR (ans = 'S') THEN
          WRITELN
      ELSE
          WRITELN(result);
    END;
END;
```

Now we have sorted out the individual components of the algorithm we can use them in the following program.

```
PROGRAM graphexample (datafile,result,INPUT,OUTPUT);
CONST  scrmaxy = 100; scrmaxx = 200;
TYPE   screen = ARRAY[0..scrmaxy,0..scrmaxx] OF CHAR;
VAR  x, intervalx, intervaly, lowbd, highbd : REAL;
     maxf, minf : REAL;
     screenwidth, screendepth : INTEGER;
     picture : screen;
     result : TEXT;
FUNCTION f(x : REAL) : REAL;
BEGIN
     f := x * x;
END;

(* place the procedures after  the  function definition
*)

BEGIN
     screensize;
     blanks;
     xscale(lowbd,highbd,intervalx);
     minmax(maxf,minf,lowbd,intervalx,screenwidth);
     intervaly := (maxf - minf) / screendepth;
     axis(highbd,lowbd,intervalx,maxf,minf,intervaly);
     graph(maxf,minf,lowbd,intervalx,intervaly);
     displayscreen;
END.
```

The result of running this program is an approximation, shown below, to the graph of $y = x*x$ for x values between -5 and 5. The parameters lowbd, highbd, screenwidth and screendepth were set at $-5, 5, 70$ and 20 respectively.

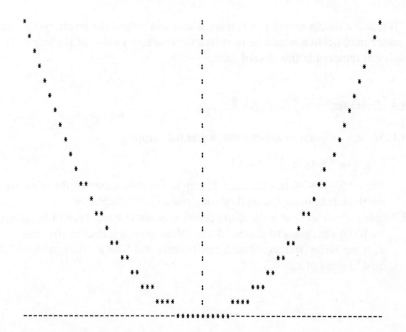

The second graph is of sin(x)/(x+1), obtained from the same program by altering the line in the function f and the range of values of x from 0 to 25. That is, lowbd:=0 and highbd:=25. The values of screendwidth and screendepth are 20 and 70 as for the previous example.

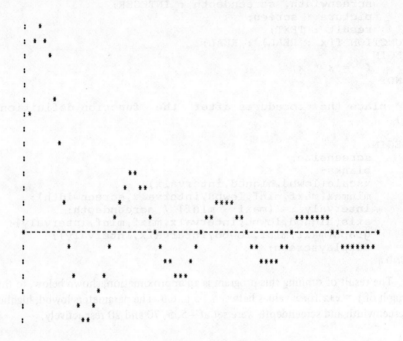

It is also a simple exercise to rotate the axis and output the screen on its side. A minor modification would be to switch the characters used for the axis if the graph was required in this altered form.

13.4 Exercises

13.1. Write a program to explore the sequential mapping

$$x_n = a\,x_{n-1}(1 - x_{n-1})$$

for $-2.5 < a < 4$. In each case, display the first 80 values of the sequence on the screen using the methods described in this chapter.

13.2. For each value of a in the above problem generate a sequence of 80 values. Write a program to display all the values in each sequence after the first ten terms, starting in each case from $x_0 = 0.5$ for a taking values -2.5 to 4 in steps of size 0.2.

13.3. Modify the graphics program to display two graphs on the same screen or page. You will need to use a different character for each graph. Use it to display the curves

$$y = 5\cos x \quad \text{and} \quad y = 1 - x$$

used in exercise **8.6** for $-6 < x < 6$.

13.4. Modify the program to handle any number of curves between one and five. Use the program to display the five polynomials used in exercise **8.2**.

14 Data Summary

14.1 Introduction

When data is collected from an experiment or a survey, or generated as the result of numerical work on a computer, it is often very extensive. We therefore need to consider ways in which to summarise a large amount of data into more manageable and easily understood forms. This summary can be in the form of a table such as a frequency distribution, a picture such as a histogram or numerical values such as the mean and standard deviation. These things are designed to give a summary of the data so that we can visualise its shape and location, which we cannot often do from the raw values.

In this chapter we consider ways of producing such summaries and the techniques for programming them in Pascal. The theory behind the statistical concepts will not be covered in detail in this text. A useful introductory text is *A Basic Course in Statistics* by Clarke and Cooke (1985).

14.2 Frequency Distributions and Histograms

Numerical data can be of two types. It can be discrete (such as number of cars owned) or continuous (such as body weight). We therefore need to consider both possibilities in the generation of a frequency distribution. However, with both types of data our objective is to summarise them into a number of classes or groups. We begin therefore by defining a type, called *classes*, which will be an array of integers into which the frequencies will be stored. The definition of such a type will be

```
TYPE classes = ARRAY[0..MaxNoOfGroups] OF INTEGER;
```

We can then declare variables, such as *frequencydistribution*, to be of this type and use them to hold frequencies calculated from our data.

Suppose the data is a set of integer values (that is, it is discrete) in an array called *data* of type *integervector* (ARRAY OF INTEGER).

```
TYPE integervector = ARRAY [1..MaxNoOfObservations]
                                        OF INTEGER;
```

176

Then the steps involved in the construction of a frequency distribution for discrete data are

Zero the frequency distribution

For each observation in the sample
        add 1 to the frequency corresponding to the value
        of that sample member

This is easy to program. However, we ought to include a check on the current observation to ensure that the value of the sample member is within the range of the frequency distribution. Failure to include such a check may result in an error caused by using a subscript outside the bounds of the type *classes*. The second step should therefore be modified to read

For each observation in the sample

IF the observation is in the limits of the array subscripts
THEN
    add 1 to the frequency corresponding to the value of that
    sample member.
ELSE
    Display an error message querying the data point

Thus a procedure such as the following could be used to construct the frequency distribution for a discrete data set.

```
PROCEDURE discretefreqdist(data : integervector;
                 noofobservations : INTEGER;
                 VAR frequencydistribution : classes);

VAR i,observation : INTEGER;

BEGIN

(* zero the frequency distribution *)
   FOR i := 0 TO MaxNoOfGroups DO
     frequencydistribution[i] := 0;

(* determine the frequencies of each value *)
   FOR observation := 1 TO noofobservations DO
     IF data[observation] IN [0..MaxNoOfGroups] THEN
        frequencydistribution[data[observation]] :=
          frequencydistribution[data[observation]] + 1
     ELSE
        WRITELN(' data out of range ')

END;
```

If the data is continuous, then the problem of summarising it in a frequency distribution is a little more involved. Examples of continuous measurements are height, weight and rainfall. However whenever we record such data, even if it consists of numerical values generated on a computer, we only record them to so many significant figures, effectively making them discrete. Amounts of rainfall, for example, are continuous quantities yet they are usually recorded to the nearest mm. Such data will, however, consist of a large number of possible values so that if we construct a frequency for each value in the data's range the resulting frequency distribution will have many values with zero or negligible frequency, resulting in virtually no improvement when compared with raw data.

Suppose that our continuous values are stored in a variable *data* whose type is declared to be *realvector* (ARRAY OF REAL).

```
TYPE realvector = ARRAY [1..MaxNoOfObservations]
                                           OF REAL;
```

To construct a frequency distribution for such data, therefore, we usually divide the interval, in which values of the data lie, into non-overlapping subintervals which we call classes. Then we determine the frequencies with which the observations fall into each of the class intervals. Thus we can use the same form of integer array of type *classes* as we used in the discrete case.

If we assume that the class intervals are all of the same width, then the number of the interval into which any observation falls is a simple transformation of the observation's value. We will need real variables *lowerbound*, to indicate the lowerbound of the first interval, *range*, to indicate the difference between this lower bound and the upper bound of the last interval and an integer variable, *noofintervals*, which will contain the number of intervals. Using this information then, the interval into which any observation lies is the largest integer not exceeding

$$(observationvalue - lowerbound)*noofintervals / range$$

which could easily be placed in a function of type INTEGER as follows

```
FUNCTION interval (observationvalue,lowerbound,
                   range   :  REAL;
                   noofintervals : INTEGER) : INTEGER;
BEGIN
   interval := TRUNC((observationvalue - lowerbound)
                                /range * noofintervals)
END;
```

Note that the first interval has the value zero which enables us to be consistent with the discrete case for handling non-negative integers.

The construction of the frequency distribution then follows the same steps as in the discrete case. A possible procedure is the following which requires the values for *lowerbound, range* and the *noofgroups* to be entered from the terminal.

```
PROCEDURE freqdisttwo(data:realvector;
                NoOfobservations : INTEGER;
                VAR  frequencydistribution : classes;
                VAR lowerbound,classinterval : REAL );

VAR i,observation,observationinterval : INTEGER;
    range,datamean,datavariance,datadeviation : REAL;

BEGIN
(* zero the frequency distribution *)
  FOR i := 0 TO MaxNoOfgroups DO
    frequencydistribution[i] := 0;

(* determine the size and boundaries of the intervals*)
  datamean := mean(data, NoOfobservations);
  datavariance := variance(data, datamean,
                                    NoOfobservations);
  datadeviation := SQRT(datavariance);
  classinterval := (4 * datadeviation) / 10;
  lowerbound := datamean - 2 * datadeviation -
                                    classinterval;
  range := datamean + 2 * datadeviation +
                        classinterval - lowerbound;

  (* determine the frequencies *)
  FOR observation := 1 TO NoOfobservations DO
    BEGIN
    (* determine the interval *)
      observationinterval:= interval(data[observation],
                            lowerbound,range,12);
    (* increment the corresponding frequency *)
      IF observationinterval < 1 THEN
          frequencydistribution[0] :=
                          frequencydistribution[0] + 1
      ELSE
          IF observationinterval > 12 THEN
             frequencydistribution[13] :=
                          frequencydistribution[13] + 1
          ELSE
          frequencydistribution[observationinterval]:=
          frequencydistribution[observationinterval] + 1
    END
END;
```

This approach of summarising data in terms of the frequencies with which observations fall into a set of class intervals that cover the range can also be used for discrete data which can take a large range of values. For example, marks in the range 0 to 100.

This leaves us with the problem of how to calculate the size of the class interval and the lower bound. One solution is to do this manually and then supply the information about the number of intervals, the lowerbound of the first interval and the length of each of the intervals, in response to prompts from the program. In many ways this is advantageous since we can choose the boundaries of the intervals to make it clear into which each data value should go. For example, if we are dealing with rainfall recorded to the nearest milli-metre, then by using half-millimetres in the boundaries of the intervals it is clear into which of the classes each of the data values will fall. We can also choose the number of classes so that a sensible summary of the data results. It is all too easy, if one value is very different in magnitude from the rest of the data, to produce a frequency distribution that has most of the data in one or two classes, making it very difficult to interpret. If the first run of the program does not produce an adequate summary, we can always revise the information supplied to the program.

We might also consider writing a procedure to make the decisions about these values. It is difficult to produce such a procedure which meets all the require-ments which might possibly be placed upon it. Almost certainly it will produce class boundaries that are not very sensible. In most statistical texts the usual recommendation is to use between 8 and 12 groups since it is felt that this gives an adequate summary of the data without too much compression. We next need to determine the boundaries of these intervals which can be summarised in terms of the lower bound and the interval length. Obvious candidates for these two values are

> lowerbound = smallest observed data value

> class width = range/(number of intervals − 1)

where range = largest observed value − smallest observed value.

However, if either of these two extreme values is a long way from the main part of the data this results in a class width that is too wide to summarise adequately the main part of the data and other class intervals that contain few or no observations.

One way of attempting to avoid this problem is the following. Suppose we use 14 intervals, and take 10 of these to cover the central part of the data, the other 4 being used to cover the tails of the distribution. Then, for most data sets, approximately 95 per cent of the data will lie in the interval

> (mean − 2 ∗ standard deviation, mean + 2 ∗ standard deviation)

We therefore take as the length of our class interval the length of this interval − that is, 4 standard deviations − divided by 10, the number of intervals used to cover it. Then the two intervals immediately on either side of these 10 are also taken to have the same length. The final two classes at each end, 0 and 13, are then used to accumulate all the observations that lie below or above the other 12 classes. This gives us the following algorithm:

> Calculate the mean and standard deviation of the observations.
> Calculate the class interval length (4 ∗ standard deviation) / 10.
> Calculate the lower bound of interval one,

that is mean − 2 * standard deviation − class length.
Calculate the range as
the upperbound of interval twelve − lower bound of interval one,
that is mean + 2 * standard deviation + class length − lower bound.

For each observation

calculate the interval in which it lies using the interval function
IF interval < 1 THEN

increment frequency for interval zero
ELSE

IF interval > 12 THEN

increment frequency for interval thirteen
ELSE

increment the frequency for that interval

If we use the functions described in section 14.3 for calculating the mean
and variance, we can use the following procedure:

```
PROCEDURE contfreqdist (data : realvector;
                noofobservations : INTEGER;
                VAR frequencydistribution : classes;
                VAR lowbound,intervalwidth : REAL;
                VAR noofgroups : INTEGER);

VAR i,observation,observationinterval : INTEGER;
    range : REAL;

BEGIN

(* zero the frequency distribution *)
  FOR i := 0 TO MaxNoOfGroups DO
    frequencydistribution[i] := 0;

(* enter necessary parameters for construction of the
    frequency values *)
  WRITELN('enter values for lowbound',
                            'range,noofgroups');
  READ(lowbound,range,noofgroups);
  intervalwidth := range/noofgroups;

(* determine the frequencies for each interval *)
  FOR observation := 1 TO noofobservations DO

    BEGIN
      observationinterval := interval(data[observation],
                            lowbd,range,noofgroups);
      IF observationinterval IN [0..noofgroups-1] THEN
        frequencydistribution[observationinterval] :=
          frequencydistribution[observationinterval] + 1
      ELSE
        WRITELN(' data out of range ')
    END
END;
```

14.2.1 Histograms

As well as presenting the data in the form of a frequency table, it is useful to produce a picture which gives an immediate impression of the data. If the data is discrete, then the usual way of illustrating this is using a spike diagram as shown in figure 14.1 for the first data set whose values are given in table 14.1.

Table 14.1 Number of radioactive particles recorded in 200 five-second intervals

0	2	4	3	5	7	5	4	3	3	1	0	2	3	4	3	7	6	2	3
4	1	2	0	3	5	0	6	3	8	3	4	2	5	2	1	2	3	1	0
2	3	0	2	3	0	3	4	2	1	2	1	3	2	4	3	6	2	3	1
6	2	3	1	0	0	2	1	2	1	3	2	2	1	3	2	5	4	0	0
3	4	2	5	3	3	2	1	0	0	2	3	2	2	3	4	1	0	5	3
0	5	0	3	2	2	1	3	0	5	4	3	0	2	3	2	4	1	0	0
0	3	2	4	1	0	2	3	1	0	3	4	6	2	2	8	1	2	1	3
0	3	4	2	3	2	3	1	2	1	3	3	2	4	3	1	0	0	2	2
2	2	1	1	1	2	1	2	4	3	1	2	5	1	0	2	0	1	2	1
2	3	0	1	9	2	1	0	3	4	2	2	3	2	2	0	2	3	3	2

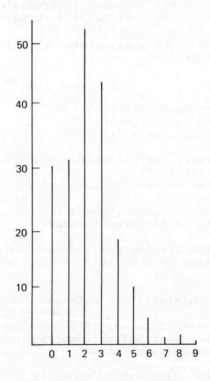

Figure 14.1 Summary of the radioactive particle data

The data set is a record of the number of radioactive particles recorded in 200 five-second intervals, and the nature of this diagram emphasises the fact that the data is discrete.

If the data is continuous or has been summarised in the form of a grouped frequency distribution, then we usually represent it in the form of a histogram. The histogram for our second data set is shown in figure 14.2. The values are the weights of 200 people and are given in table 14.2.

Table 14.2 Weights in lbs of 200 people

181	183	179	170	173	165	182	168	176	161
171	177	189	178	184	173	168	177	183	187
183	170	191	183	170	179	176	164	180	176
183	165	178	182	170	170	191	189	185	178
187	181	172	170	168	181	172	181	188	177
180	175	172	189	175	172	182	175	161	178
173	170	179	171	174	170	171	174	179	163
166	162	187	174	189	164	173	180	179	171
173	184	173	171	181	162	168	179	181	178
191	173	172	174	185	187	178	170	177	163
183	179	179	180	183	188	168	178	181	175
173	173	177	176	176	164	169	174	181	180
182	182	179	185	164	176	181	194	162	177
178	183	180	175	169	169	175	177	180	174
178	175	186	175	176	186	179	177	176	167
169	171	177	179	166	171	173	177	190	172
178	182	179	167	184	182	175	173	173	167
189	179	150	184	166	188	165	177	175	182
168	165	184	167	192	183	186	167	170	186
164	170	181	184	155	177	178	181	159	192

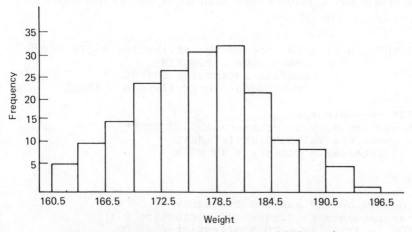

Figure 14.2 Histogram of the weights of 200 people

From the picture we see that the frequency in each class is represented by a rectangle using the class interval as the base. The areas of the rectangles represent the frequencies, although in examples such as ours (where the class intervals are all equal) their heights are also proportional to the frequencies.

A crude representation of such a diagram can be obtained on a screen or lineprinter.

Suppose we have a frequency distribution stored in a variable *frequencydistribution* using the methods described in the previous section. We can represent the frequency of each group by a line of stars (say), where the length of the line represents the frequency. Obviously, if the frequency in any interval is larger than the number of character positions available we must scale down the frequencies in order to accommodate the largest. The steps in a possible algorithm to determine the maximum frequency are

IF the maximum frequency > no. of characters per line THEN
              scale down all the frequencies
FOR each interval
        display numerical information.
                    For example lower bound, upperbound, frequency
            Determine the scaled down frequency for the interval
            FOR j:= 1 to scaled down frequency for the interval
                        display a star
        move to the next line

We can now translate this into a procedure. To accommodate both discrete and continuous data we use a BOOLEAN parameter *continuous* which has the value TRUE if the data is continuous and FALSE if it is discrete. This is used to control the display of the upper boundary of the class interval since this is only required in the continuous case. We also have an INTEGER parameter *nooflastinterval* which indicates how much of the array *frequencydistribution* we wish to display. The other parameters are self-explanatory. The result is a procedure such as the following:

```
PROCEDURE histogram (frequencydistribution : classes;
                continuous : BOOLEAN;
                nooflastinterval : INTEGER;
                lowerbound, intervalwidth : REAL);

CONST maxwidth = 60;
VAR maxfrequency,lowerintervalbound : REAL;
    upperintervalbound,scale :REAL;
    j,interval,noofstars : INTEGER;

BEGIN

(* determine the maximum frequency *)
  maxfrequency := frequencydistribution[0];
  FOR interval := 1 TO nooflastinterval DO
```

```
  IF maxfrequency < frequencydistribution[interval]
                                                 THEN
        maxfrequency := frequencydistribution[interval];

(* determine the scaling factor *)
   scale := 1;
   IF maxfrequency > maxwidth THEN
         scale := maxwidth / maxfrequency;

(* determine the boundaries of the first interval *)
   lowerintervalbound := lowerbound;
   upperintervalbound := lowerbound + intervalwidth;

(* display the numerical information and the crude
histogram *)
   FOR interval := 0 TO nooflastinterval DO

     BEGIN
       IF continuous AND (interval = 0) THEN
          WRITE(resultsfile,'    under ')
       ELSE
          WRITE(resultsfile,lowerintervalbound:9:2);
       IF continuous THEN
          IF (interval < nooflastinterval) THEN
             WRITE(resultsfile,upperintervalbound:9:2)
          ELSE
             WRITE(resultsfile,' and over');
       WRITE(frequencydistribution[interval]:4,' ':4);
       noofstars :=
          ROUND(frequencydistribution(interval)*scale);
       FOR j := 1 TO noofstars   DO
            WRITE('*');
       WRITELN;
       lowerintervalbound := lowerintervalbound +
                                          intervalwidth;
       upperintervalbound := upperintervalbound +
                                          intervalwidth
     END;
END;
```

14.2.2 Examples

To complete this section we consider various examples of calls to these procedures using our two example data sets.

If we are using a maximum width of display of 60 characters for the histogram, then for the discrete data of table 14.1, calling the procedures with parameters as shown in program *discretefreqtest* where *x* is of type *integervector*, we get the display in figure 14.3. The text of the procedures is given earlier in the chapter; only the headings are repeated so that the values of parameters used can be clearly seen.

```
PROGRAM discretefreqtest (datafile,INPUT,OUTPUT);
CONST MaxNoOfgroups = 15;
TYPE classes = ARRAY[0..MaxNoOfgroups] OF INTEGER;
     integervector = ARRAY [1..200] of INTEGER;
VAR x : integervector;
    xfreq : classes;
    i,noofobservations : INTEGER;
    lowbd,intervalwidth : REAL ;
    datafile : TEXT;

PROCEDURE discretefreqdist(data : integervector;
                  noofobservations : INTEGER;
                  VAR frequencydistribution : classes);
(*   procedure  to  determine  a  discrete  frequency
distribution *)

PROCEDURE histogram (frequency : classes;
                  continuous : BOOLEAN;
                  nooflastinterval : INTEGER;
                  lowerbound, intervalwidth : REAL);
(* procedure to display histogram  on a line printer or
screen *)

BEGIN
   RESET(datafile);
   READ(datafile,noofobservations);
   FOR i := 1 to noofobservations DO
      READ(datafile,x[i]);
   discretefreqdist(x,noofobservations,xfreq);
   histogram(xfreq,FALSE,10,0,1);
   WRITELN;
END.
```

```
     0.00   31   *********************************
     1.00   32   *********************************
     2.00   54   ***********************************************************
     3.00   45   **************************************************
     4.00   18   ******************
     5.00   10   **********
     6.00    5   *****
     7.00    2   **
     8.00    2   **
     9.00    1   *
    10.00    0
```

Figure 14.3 Summary of the radioactive particle data

The effect of the scaling is shown in figure 14.4 which shows the same data
constructed with a maximumwidth of 25 characters for display of the histo-

gram. Note how the relative shape is still retained, although very small values such as 1 do not appear since they scale down to zero.

```
 0.00   31   ***************
 1.00   32   ***************
 2.00   54   *************************
 3.00   45   **********************
 4.00   18   ********
 5.00   10   *****
 6.00    5   **
 7.00    2   *
 8.00    2   *
 9.00    1
10.00    0
```

Figure 14.4   Summary of the radioactive particle data – scaled histogram

Results from using both procedures for continuous data are illustrated in figures 14.5 and 14.6.

```
  under   153.50    1   *
 153.50   157.50    1   *
 157.50   161.50    3   ***
 161.50   165.50   14   **************
 165.50   169.50   18   ******************
 169.50   173.50   36   ************************************
 173.50   177.50   36   ************************************
 177.50   181.50   42   ******************************************
 181.50   185.50   26   **************************
 185.50   189.50   16   ****************
 189.50   193.50    6   ******
 193.50  and over   1   *
```

Figure 14.5   Summary of weights using user-defined intervals

Figure 14.5 illustrates the result from using the following information supplied by the user:

lowerbound := 149.5   range := 48   noofgroups := 12

which gives us twelve intervals of length 4. The boundaries of the intervals fall between the rounded values of the recorded data, ensuring that there is no conflict about into which interval each of the recorded values falls.

```
under      160.78    3    ***
160.78    163.86     7    *******
163.86    166.94    12    ************
166.94    170.02    26    **************************
170.02    173.10    25    *************************
173.10    176.18    24    ************************
176.18    179.26    36    ************************************
179.26    182.34    26    **************************
182.34    185.42    18    ******************
185.42    188.50    11    ***********
188.50    191.58     9    *********
191.58    194.66     3    ***
194.66    197.74     0
197.74 and over      0
```

Figure 14.6   Summary of the weights using intervals chosen by the program

Figure 14.6 shows the result of using the suggested method for automatic determination of the interval boundaries.

Notice how this automatic method produces intervals of length 3.08, which would not be chosen naturally, and perhaps a representation using an interval width of length 3 should be produced using the first procedure.

14.3 Moments

The frequency distribution summarises the data in the form of a table. We can also draw a picture, such as a histogram, to illustrate the general shape of the data. The main features which describe a data set are its location or 'centre' and the way in which the data is distributed about this 'centre'. This distribution around the centre can generally be described in terms of its spread and its skewness. As well as representing these features pictorially, we can also represent them numerically in terms of what are called moments by analogy with the use of the term in applied mathematics and physics. In this section, therefore, we consider the calculation of a handful of numerical values which summarise these features of the data. We describe ways of writing programs to estimate the centre and spread, leaving the higher-order moments to the reader as an exercise.

14.3.1   Mean and median

We have already met the mean or average as an example to illustrate other features of the Pascal language. It is simply the sum of the observations divided

by the number of observations. Thus, if the data is available in an array, called
data of type *realvector* described earlier, then a simple function to calculate
the mean of these observations is

```
FUNCTION mean(data : realvector;
                noofobservations : INTEGER) : REAL;
VAR i : INTEGER;   sum : REAL;
BEGIN
   sum := 0;
   FOR i := 1 TO noofobservations DO
      sum := sum + data[i];
   mean := sum / noofobservations;
END;
```

If the data is summarised in terms of a frequency distribution consisting of
frequencies f_i corresponding to interval i, then the mean is given by

$$\bar{x} = \frac{\Sigma x_i f_i}{\Sigma f_i}$$

where the x_i values used are the mid-points of the class intervals. This approximation has to be made since the information about the individual x values is
assumed to be unavailable once we have summarised the data into a frequency
distribution. In using the mid-point we are assuming the data in each interval to
be equally spread throughout the interval, which turns out to be a reasonable
approximation. The calculation in this case is a little more involved. If we
assume *lowerbound* is a REAL variable representing the lower bound of the
frequency distribution, *intervallength* is a REAL variable representing the
length of each class and *frequencydistribution* is a variable of type *classes*
which contain the frequencies, then the calculation can be performed as
follows:

```
midpoint := lowerbound + intervallength / 2;
sum : = 0;
noofobservations := 0;
FOR j := 0 TO noofintervals-1 DO
   BEGIN
     sum := sum + midpoint * frequencydistribution[j];
     noofobservations := noofobservations +
                           frequencydistribution[j];
     midpoint := midpoint + intervallength
   END;
mean := sum / noobservations;
```

The mean is the most commonly used of the measures of location since it is
easy to manipulate accurately and easy to calculate. It does, however, suffer
from the disadvantage that if the data is skew or has one or two very large or
small values, it may not give a good representation of the data. We therefore
also consider an alternative measure of the 'centre' of a set of data.

An alternative measure of location that we can use is the median, which is defined as the value which divides the data into two equally numerous sets. Thus, for an odd number of observations it is the value of the middle observation when the data has been placed in order of magnitude. If the number of observations is even, it is usually taken to be the average of the middle two. The problem of sorting data is one that is covered by most books on programming and hence is not covered in detail in this text. However, a procedure for sorting data is given in chapter 18.

Using this procedure, if we have a vector *data* of real observations which is sorted into a vector *sorteddata*, then the median is calculated as follows:

```
sort(data,sorteddata,nofobservations);
IF ODD(noofobservations) THEN
   median := sorteddata[(noofobservations DIV 2) +1]
ELSE
   median := (sorteddata[noofobservations DIV 2] +
        sorteddata[(noofobservations DIV 2) +1]) / 2;
```

If the data is summarised in the form of a frequency distribution, then it is already effectively sorted. In this case we determine the median class, which is the class for which the cumulative frequency becomes greater than half the total frequency. If the frequency distribution represents discrete values, then the median is simply the value of this class. However, if the data is grouped we assume that the data that falls into that group is equally spaced throughout that interval and perform a linear interpolation. Thus the median can be calculated using the formula

$$\text{median} = \text{lbd} + \frac{c * ((n + 1)/2 - f_c)}{f_m}$$

where lbd = lower bound of the interval in which the median lies

$\quad c$ = width of this interval

$\quad f_c$ = the cumulative frequency up to the lower boundary of this interval

$\quad f_m$ = frequency in the interval in which the median lies

$\quad n$ = the total number of observations

The following program steps perform the necessary calculations.

```
halffrequency := (totalfrequency + 1) / 2;
cumulativefrequency := 0;
medianinterval := 0;

(* find which is the median class *)
WHILE cumulativefrequency + frequency[medianinterval]
                                    < halffrequency DO
   BEGIN
   cumulativefrequency := cumulativefrequency +
                            frequency[medianinterval];
   medianinterval := medianinterval + 1;
   END;
```

```
(* calculate the median using linear interpolation
   note:    distributionlowerbound  is the lowerbound of
            the  range  covered by the  whole  frequency
            distribution *)
   lowerendofmedianclass := distributionlowerbound +
                    (medianinterval - 1 ) * intervallength;
   median := lowerendofmedianclass + intervallength *
      (halffrequency - cumulativefrequency) /
                              frequency[medianinterval];
```

14.3.2 Measures of dispersion

The most commonly used measure of dispersion or spread of a data set is the
standard deviation. This is the square root of the variance which is defined as the
sum of the squares of the deviations of the observation values from their mean,
divided by the number of observations. Thus, if S is the standard deviation,
then we have

$$S^2 = \frac{1}{n} \, \Sigma(x_i - \overline{x})^2$$

For small samples, the sum of squares is usually divided by $n-1$ for reasons
beyond the scope of this text (see Clarke and Cooke (1985) for a detailed
discussion). However, for large samples there is little difference since n and
$n-1$ are virtually the same, and we develop all programs in terms of the formula
given.

If we examine this sum, we see that observations that are close to the mean
have a small value for $x_i - \overline{x}$, while for observations that are far from the mean
this value is much larger. Thus, if the data is concentrated near the mean then
we can expect the value of S to be fairly small, while if the data is well spread
out this value will be large. Thus the standard deviation S gives us a measure of
how much the data is spread out.

It is generally found that approximately 95 per cent of the observations lie
within 2 standard deviations of the mean.

The variance can be calculated using the formula given as follows:

```
FUNCTION variance(data : vector;
                  noofobservations : INTEGER;
                  datamean :REAL) : REAL;
(* function to calculate the  variance  using  a
previously calculated  mean *)

VAR sum : REAL; j : INTEGER;

BEGIN
  sum := 0;
  FOR j := 1 TO noofobservations DO
    sum := sum + (data[j] - datamean) *
                          (data[j] - datamean);
  variance := sum / noofobservations;
END;
```

The formula given can also be shown to be equal to

$$S^2 = \frac{1}{n} \sum x_i^2 - \bar{x}^2$$

$$= \frac{1}{n} \sum x_i^2 - \left(\frac{\sum x_i}{n}\right)^2$$

and in this form it can be used to calculate the mean and variance in a single loop as follows:

```
sum := 0;
sumofsquares := 0;
FOR j := 1 TO noofobservations DO
   BEGIN
      sum := sum + data[j];
      sumofsquares := sumofsquares + data[j] * data[j]
   END;
mean := sum / noofobservations;
variance := sumofsquares / noofobservations -
                                         mean * mean;
```

While this appears a more efficient method for performing the calculations, and in many situations will be adequate, it is more prone to numerical errors than the alternative. If all the data consists of very large values that differ little from one another, then the two quantities involved in the final subtraction will be very similar, producing a large error in the operation.

The formula for calculating the variance of a frequency distribution is analagous to that for the mean. It is

$$S^2 = \frac{\sum f_i (x_i - \bar{x})^2}{\sum f_i}$$

We will leave the writing of a procedure to perform such an operation as an exercise for the reader.

14.4 Exercises

14.1. The following values are the times between the arrival of successive calls at a telephone exchange in sections:

```
2.6  0.6  1.6  4.9 36.1  1.2  5.7 27.8  6.8  4.6  0.7 5.7 9.6  3.8 8.4
1.0 39.1 27.7 12.9  9.2 39.0 22.3  3.0 16.8  1.2 14.1 8.1 1.7 11.2 3.0
3.1  1.4  9.7  0.1 20.6  2.9  0.0  0.1  3.3 21.8  4.2 7.7 3.0  1.2 0.8
45.6  1.0 11.7  4.0  2.9  5.4  6.8  0.3 15.2 17.2  8.5 9.5 4.2  0.4 6.5
```

Write a program that reads the data into a vector, and calculates its mean and variance. Summarise the data in a frequency distribution and produce a crude histogram of the data.

14.2. Write a function that calculates the variance, given the data in the form of a frequency distribution.

14.3. The geometric mean of n values $x_1, x_2 \ldots x_n$ is defined by $g = (x_1 x_2 \ldots x_n)^{1/n}$. Write a function to calculate this value for a given vector of observations.

Explain why the use of the result

$$\log g = (1/n) \sum_{i=1}^{n} \log(x_i)$$

may be the best way to perform this calculation.

14.4. The mean and variance are often called the first two moments of a distribution. In general, the rth moment about the origin is defined by

$$m'_r = \Sigma f_i x_i / \Sigma f_i$$

while the rth moment about the mean is defined by

$$m_r = \Sigma (x_i - \overline{x}) f_i / \Sigma f_i$$

Thus the mean x is m'_1 while the variance is m_2.

Write a function to calculate the rth moment about the origin and the rth moment about the mean, and calculate the first four moments in each case for the above data set.

Two other quantities that are used to summarise the shape of a frequency distribution are the skewness defined by

$$\beta_1 = m_3/m_2$$

and the kurtosis defined by

$$\beta_2 = m_4/m_2^2$$

Calculate these values for the above data set.

14.5. Given a sample of n pairs of observations (x_i, y_i) $i = 1 \ldots n$ on two variables X and Y, the product moment coefficient which measures the strength of relationship between X and Y is defined by

$$r = \frac{\Sigma (x_i - \overline{x})(y_i - \overline{y})}{\Sigma(x_i - \overline{x})^2 \; \Sigma(y_i - \overline{y})^2}$$

Write a function to calculate r, given that the data is stored in two vectors of length n.

14.6. Given a sample of n pairs of observations (x_i, y_i) $i = 1 \ldots n$ on two variables X and Y, the least squares estimates of the parameters α and β in the regression line $y = \alpha + \beta x$ are given by

$$b = \frac{\Sigma (x_i - \overline{x})(y_i - \overline{y})}{\Sigma (x_i - \overline{x})^2}$$

and $a = \overline{y} - b\overline{x}$.

Write a procedure to calculate these estimates.

14.7. Write a further procedure to calculate the residuals $e_i = y_i - a - b\,x_i$ and which indicates any values of e_i which are not within two standard deviations of the mean error (that is, zero).

   Use the character graphics program to display the residuals and the lines 2 standard deviations above and below zero.

14.8. Write a program that summarises the dates on which Easter Sunday occurs between 1582 and 4903 in a frequency distribution. (Use the algorithm described in exercise **3.9**.)

14.9. Modify the above program so that it produces a list of the years when the extreme dates occur.

14.10. The mode is the most frequently occurring value or the group with the largest frequency. If the data is discrete and the frequencies correspond to values of the original data then the mode is simply the value with the maximum frequency. Write a function to find the mode of a discrete set of data.

   However, if the data is continuous or has been grouped, then we can either quote the modal class or use an interpolation on the basis that the mode is closer to the neighbouring group which has the larger frequency. If the distribution is not J-shaped, then the mode is calculated by the formula

$$\text{mode} = \text{lbd} + \frac{c * (f_m - f_{m-1})}{2*f_m - f_{m-1} - f_{m+1}}$$

where lbd = lower bound of modal interval
$\quad\quad c$ = length of the class interval
$\quad\quad m$ = number of the modal interval
$\quad\quad f_m$ = frequency of modal interval
$\quad f_{m-1}$ = frequency of previous interval
$\quad f_{m+1}$ = frequency of next interval

Write a function to calculate the mode in this case.

15   Random Numbers

15.1   Random and Pseudo-random Numbers

Random numbers are used in the generation of values from specific distributions. These are used in Monte-Carlo simulation methods which are used extensively in work on stochastic (probabilistic) problems such as queueing, spread of disease and epidemics, stock and inventory control, floodwater control and dam construction.

Somewhat surprisingly, simulation methods can also be used to obtain numerical solutions to certain non-probabilistic problems that do not easily yield to more conventional mathematical and numerical attack. They have been used in applications that are both probabilistic and deterministic in nature. W. S. Gossett, more commonly known under his pseudonym, Student, in 1908 used random numbers in simulating the distribution of a correlation coefficient and later in his work on Student's t distribution. From the results of simulations he was able to postulate certain results about these two distributions which were later proved theoretically.

In a later chapter we will consider some elementary simulation examples related to queueing situations, but here we concentrate on ways of attempting to produce random numbers and on how to test whether the resulting numbers are adequate. We also consider some simple methods for generating random samples from some of the more common statistical distributions such as the Exponential, Normal and Binomial. An excellent and very readable introductory text on the subject of random numbers and simulation in general is *The Elements of Simulation* by Morgan (1984).

A set of random numbers can be thought of as a set of numbers which have no discernible pattern nor can they be generated according to some rule. Thus a set of numbers can be considered to be random when

> "No prediction of the nature of forthcoming values is possible from the values already available"

This of course immediately poses us a problem when we come to consider random numbers in a programming context. Before the age of the computer, people engaged in simulation used to make use of tables of pseudo-random numbers generated by some mechanism. The compilations of random numbers by Tippett (1927), who used digits read from log tables, and Kendall and

Babbington-Smith (1939), who used rotating discs stopped at random, being the most well known. However, to store such compilations, which contained 10 000 and 100 000 entries respectively, would present us with serious problems when we turn to using computers. There would also be many applications for which the number of random numbers would not be sufficient. We therefore use pseudo-random numbers which are a sequence of numbers generated by a sequential mapping but give every appearance of being random.

Because pseudo-random numbers are numbers generated by some mechanism which is known, they suffer from an obvious disadvantage. The numbers produced are predictable in the sense that, given sufficient information regarding the preceding numbers, the next one can be predicted exactly. However, it is possible to generate sequences which can be considered to be random since they apparently satisfy all other criteria that we feel such numbers should satisfy.

Since on a digital computer we are restricted to finite number sets, while it is possible to generate infinite sequences of numbers they will contain cycles of finite length that are repeated. Our problem is therefore constrained by the need to find a method for generating numbers whose cycle is long enough so that, within the context of the problem on which we are working, we do not use more numbers than are available in a single period of this cycle.

There are therefore two things that we need to consider. Firstly the methods that we can use to generate a suitable sequence and secondly the properties that we expect it to have and methods for checking whether it possesses these properties.

15.1.1 *Generation of pseudo-random numbers*

The earliest method for generating such a sequence was the mid-square method. This consisted of taking a random number, called the 'seed', of n digits, squaring it and taking the middle n digits of the resulting $2n$ digit number as the next number. This posed problems about choosing a random seed and avoiding a seed that generated only short sequences.

However, while not all starting values produce chains long enough to be useful, it is possible to find some values that generate sets of numbers which are long enough to merit consideration.

An obvious extension of this is to start with two numbers and use their product as the next number in the pseudo-random sequence. Each succeeding term in the sequence is then the product of the previous two terms. This, though, suffers from similar problems to the mid-square technique.

The method which evolved out of these ideas and is now firmly established as a method of generating pseudo-random numbers is the linear mixed congruential method. The method consists of starting with a pseudo-random number U_k and generating the next number U_{k+1} according to the rule

$$U_{k+1} = (a\,U_k + c) \text{ modulo } m$$

That is, the next number in the sequence is the remainder from dividing $a U_k + c$ by m. The values of a, c and m are chosen so that the sequence of numbers generated is as long as possible without being repeated. The values U_k/m resulting from this are used as a pseudo-random sequence of values in the interval [0,1]. They are assumed to be equally likely to have come from anywhere in this interval and hence represent a set of values from what is called a uniform distribution.

Clearly, such a method of generation can produce no more than m numbers at most before they start repeating themselves. Since most computers perform arithmetic to base 2, m should be of the form 2^k where k is such that the numbers generated are in the range allowed on our particular computer.

It can be shown that a sequence of numbers generated in this way has full period — that is, all m values occur before the cycle repeats itself if a, c and m satisfy the following three conditions:

(1) c and m have no common divisors.
(2) $a = 1 \pmod p$ for every prime factor p of m.
(3) $a = 1 \pmod 4$ if m is a multiple of 4.

If $m = 2^k$, then (3) implies that a must be of the form $4c + 1$ for any c, and such an a also satisfies (2). Condition (1) is satisfied if c is any odd positive integer.

Not all sequences generated by such a mechanism are suitable as random sequences even if their cycle is of full length. Many sequences can, for example, be shown to have some form of correlated structure. Hence all sequences generated by a pseudo-random mechanism should be checked to see if they satisfy the properties of randomness.

Many authors have explored possible values. Ripley (1983), for example, has examined various suggestions as to appropriate values for a, c and m based on a series of far-reaching tests relating to the behaviour of the distribution of pairs, triples and quadruples of succeeding values. Among those that he considers are

$$
\begin{array}{llll}
m = 2^{32} & a = 69\,069 & c = \text{odd} \\
m = 2^{16} & a = 257 & c = 41 \\
m = 2^{16} & a = 293 & c = \text{odd}
\end{array}
$$

(The size of m is clearly restricted by the maximum size of integer that our computer can handle.)

A function to calculate pseudo-random numbers is easy to write since it involves only the results of simple assignments. The values of a, c and m are given as constants which can easily be changed should the need arise later. The values of U_k are placed in a global variable *seed* of type INTEGER so that they are not lost between calls of the function. An initial value for *seed*, that is U_0, should be assigned in the main part of the program before the function *random*

is first used. This approach avoids using a variable parameter which, while
allowed in the case of Pascal functions, is contrary to the usual idea of a function
returning only a single value. A suitable function is therefore

```
FUNCTION random : REAL;
CONST
    a = 293;     c = 1;      m = 65536;
BEGIN
   seed := (a * seed + c) MOD m;
   random := seed / m;
END;
```

which generates random values over the interval $[0,1]$.

One advantage that pseudo-random numbers have is that they are repeatable,
which means that we can consider various models under the same conditions of
simulation without having to store our values in any way. Starting from the same
initial value or seed each time, we will generate the same sequence of values.

15.2   Tests for Randomness

Sequences of pseudo-random numbers can be tested to see if, subject to the
limitations of their derivation, they satisfy the properties we expect of such a
sequence. The fundamental properties that we expect of such a sequence are
uniformity and *independence*. The values are assumed to come from a uniform
distribution over the interval $[0,1]$, hence we should expect them to be evenly
spread out over that interval. Thus the first test we consider is a goodness-of-fit
test of such a proposition.

15.2.1   Goodness-of-fit test of uniformity

If we divide the interval $[0,1]$ into N non-overlapping subintervals of equal
length, then we would expect that values in each subinterval should occur with
equal frequency. The way to carry out such a test is to count the number of
times that values fall in each subinterval, then perform a test using the χ^2 distribution to support or reject this hypothesis.

Thus the random numbers generated are a data set and we simply construct
a frequency distribution using the techniques discussed in chapter 14. These are
called the observed frequencies which we represent by O_i ($i = 0 \ldots N-1$). If the
generated numbers were from a uniform distribution, then the expected frequencies E_i ($i = 0 \ldots N-1$) are the number of random numbers generated, n,
divided by the number of intervals, N.

Then the value $\Sigma (O_i - E_i)^2 / E_i$ has a χ^2 distribution with $N-1$ degrees of
freedom. Obviously, if the observed values differ greatly from the expected

values, then the sum will have a large value. Hence we reject the hypothesis that the values are suitably uniform if we observe a sufficiently large value of the sum. Decisions as to how to interpret the phrase 'sufficiently large' and the full details of the test are given in Clarke and Cooke (1985) and other introductory statistical texts.

Using the function described earlier to generate random numbers and the procedures in section 14.2, figure 15.1 shows the histogram obtained from 200 numbers starting from the seed 6097. The resulting observed χ^2 value is 6.80 which is not significant. Hence this part of the sequence results in a sufficiently uniform spread of values.

```
0.00   0.10   24   ************************
0.10   0.20   20   ********************
0.20   0.30   16   ****************
0.30   0.40   14   **************
0.40   0.50   19   *******************
0.50   0.60   27   ***************************
0.60   0.70   17   *****************
0.70   0.80   20   ********************
0.80   0.90   20   ********************
0.90   1.00   23   ***********************
```

Figure 15.1   Frequency distribution for 200 random numbers

15.2.2   Serial tests

Successive values in a sequence of random numbers should be uncorrelated with one another. Thus, if we consider non-overlapping pairs of values $(U_1/m, U_2/m)$, $(U_3/m, U_4/m) \ldots$, then these points should occur evenly over a unit square. We can therefore divide this square into subsquares or cells by dividing the range of values for each of the numbers in the pair into non-overlapping intervals of equal length. We then record the number of pairs of observations that lie in each cell. If the sequence was random we would expect each cell to contain an equal number of observations. We can again test this using a χ^2 goodness-of-fit test. Table 15.1 shows the resulting frequencies using our function *random* starting from a seed of 697. This has an observed χ^2 value of 86.00 which is not significant. The table of frequencies can be constructed using the methods described in section 14.2 and a two-dimensional array in which to store the frequencies.

This could be extended to threes and fours etc. along the lines of the lattice tests discussed by Ripley (1983), but this takes us beyond the realms of this text.

A second possibility is to calculate the correlation coefficient between neighbouring values. We can use

Table 15.1   Frequencies of pairs of random numbers

Upper boundary first number	Second number									
	0.1	0.2	0.3	0.4	0.5	0.6	0.7	0.8	0.9	1.0
0.1	9	12	3	5	8	7	4	6	4	10
0.2	9	9	7	12	7	10	6	6	7	7
0.3	7	4	4	5	8	9	5	8	9	4
0.4	13	4	10	11	9	5	3	5	3	8
0.5	7	7	7	8	5	6	12	8	9	9
0.6	9	11	7	7	11	13	8	7	11	7
0.7	14	4	8	7	12	8	6	6	4	8
0.8	5	9	6	4	7	5	6	6	11	6
0.9	5	7	7	4	9	5	8	6	8	11
1.0	5	7	6	10	8	9	8	10	11	13

$$\rho_1 = \frac{\sum\limits_{i=1}^{n-1} (U_i - \bar{U})(U_{i+1} - \bar{U})}{\sum\limits_{i=1}^{n} (U_i - \bar{U})^2}$$

to estimate this correlation. More generally, we can estimate the correlation between pairs of values k observations apart using

$$\rho_k = \frac{\sum\limits_{i=1}^{n-k} (U_i - \bar{U})(U_{i+k} - \bar{U})}{\sum\limits_{i=1}^{n} (U_i - \bar{U})^2}$$

It has been shown that for random sequences the mean and variance of this quantity are 0 and approximately $1/n$ respectively, where n is the number of observations considered. Hence, applying the standard test of hypothesis, if we observe values of ρ_k which are within $1.96/\sqrt{n}$ of zero then our sequence can be considered to be random. Table 15.2 gives the first 20 such correlations for 1500 numbers generated using our function *random* starting from a seed of 697 – none of which is outside the interval $(-1.96/\sqrt{n}, 1.96/\sqrt{n})$.

Table 15.2   Values of ρ_k for $k = 1 \ldots 20$ for the first 1500 numbers generated by the function *random*

0.05	0.04	0.01	0.01	−0.00	0.01	0.01	−0.01	0.00	−0.03
−0.05	−0.05	−0.01	0.00	0.00	0.01	−0.01	0.02	0.02	0.02

15.3 Random Numbers from other Distributions

Pseudo-random number generators give us a way of simulating values from a
uniform distribution over the interval [0,1]. In many applications we need
simulated values from a whole range of other statistical distributions such as
the exponential, normal, binomial and Poisson distributions. We therefore conclude this chapter by considering ways of converting our uniform random
variable to one from other distributions. Here we make only one or two
suggestions; details of other solutions for these and other distributions are
covered in Morgan (1984).

The following result shows that converting uniform values is, in principle,
possible although it only gives us a feasible solution in a few instances. The
result states the following.

If U is a uniform random variable over the interval (0,1), then any other
random variable X for which the cumulative probability function is

$$\text{Prob}(X < x) = F(x)$$

can be obtained from the uniform random variable using

$$X = F^{-1}(U).$$

The problem in applying such a result is that the function F must exist in
closed form and have a unique inverse. Thus, for distributions such as the
normal distribution it does not allow us to proceed but for the exponential
distribution it makes the simulation of values very easy.

15.3.1 *Continuous distributions*

The exponential distribution has density function

$$f(x) = \lambda \exp(-\lambda x) \quad \text{for } \lambda > 0 \text{ and } x > 0$$

and cumulative distribution

$$F(x) = \int_0^x \lambda \exp(-\lambda t) \, dt = 1 - \exp(-\lambda x)$$

Hence the inverse of the cumulative distribution function gives us

$$X = -(1/\lambda)\log_e(1 - U)$$

as the formula for generating values from this distribution.

Since, if U has a uniform distribution over (0,1) then so has $1-U$, we can
instead carry out the simulation using

$$X = -(1/\lambda)\log_e U$$

thereby removing one operation from the calculation of each value from the distribution.

If we want to simulate a value from a normal distribution with density

$$f(x) = \frac{1}{\sqrt{(2\pi)}} \exp(-x^2/2)$$

then we have to resort to other methods.

One approach is to make use of the *central limit theorem* which states that if we take the mean of a set of n observations from any distribution, then the distribution of the mean tends to the normal distribution as the number of observations n tends to ∞. Thus, for large sample sizes we can approximate the distribution of the mean by the normal distribution. If the original distribution is symmetric, then the value of n at which the approximation can be made is fairly small.

Thus, if we sample values from a uniform distribution over [0,1] and calculate

$$X = \sum_{i=1}^{n} U_i$$

then for a reasonable value of n, 12 say, the resulting observation has an approximate normal distribution. It is obvious from the symmetry of the distribution that a uniform random variable has a mean of $1/2$. It can also be shown that it has a variance of $1/12$. Thus our value X, obtained by summing n independent such variables, will have mean $n/2$ and variance $n/12$. Thus we can assume that our variable has an approximate $N(n/2, n/12)$ distribution. Hence

$$y = \frac{(x-n/2)}{\sqrt{(n/12)}}$$

has an approximate normal distribution with mean 0 and variance 1.

The obvious problem with such a method is that since $0 < U_i < 1$, then $0 < X < n$. That is, the resulting values are in fact bounded, hence the tails of the distribution will not be completely accurate. However, with $n = 12$ the y values will lie between -6 and 6 with a standard deviation of 1, hence, unless the number of observations used is very large, this will not cause many problems.

Other methods have been developed which are quicker and do improve the representation of the tails of the distribution. The Box–Müller method (1958) generates a pair of independent normal (0,1) random variables from a pair of independent uniform [0,1] random variables, U_1 and U_2, using the equations

$$x_1 = (-2 \log_e U_1)^{1/2} \cos(2\pi U_2)$$

$$x_2 = (-2 \log_e U_1)^{1/2} \sin(2\pi U_2)$$

This result follows from the fact that if we transform x_1 and x_2 into polar co-ordinates such that

$$x_1 = R \cos \omega \quad \text{and } x_2 = R \sin \omega$$

then it can be shown that if ω has a uniform distribution over $(0,2\pi)$, and R^2 has an exponential distribution with the parameter $\lambda = 1/2$, then x_1 and x_2 have independent normal distributions with mean zero and variance one. Hence, given U_1 and U_2 to calculate a value of ω we use $2\pi U_2$, while a value for R is given by $(-2\log_e U_1)^{1/2}$.

15.3.2 *Discrete distributions*

A value of a random variable from a binomial distribution with parameters n and p whose probabilities are described by the function

$$P(X=j) = \binom{n}{j} p^j (1-p)^{n-j} \quad j = 0,1,2\ldots n \quad 0 < p < 1$$

can be generated from a uniform random variable using the fact that the binomial random variable can be regarded as the sum of n independent random variables, each of which can take the value 1 with probability p and 0 with probability $1-p$. Thus, if

$$y_i = \begin{cases} 1 \text{ with probability } p \\ 0 \text{ with probability } 1-p \end{cases}$$

then $X = \Sigma y_i$ has a binomial distribution with parameters n and p.

A random variable of type y can be simulated very simply from a uniform distribution by setting

$$y_i = 1 \quad \text{if} \quad U_i \leqslant p$$

and $\qquad y_i = 0 \quad \text{if} \quad U_i > p$

Thus, to simulate a binomial random variable we need to simulate n uniform random variables which for small or moderate sized n will not cause too many problems. However, as n increases we have to look for other methods in order to speed the process up. The following lines of program can be used to simulate a single value from a binomial distribution with parameters n and p.

```
x := 0;
FOR i := 1 TO n DO
    BEGIN
       y := random;
       IF y < p THEN x := x + 1
    END;
```

The Poisson distribution whose probabilities are described by the function

$$P(X=j) = \frac{\lambda^j}{j!} \exp(-\lambda) \quad j = 0,1,2\ldots \lambda > 0$$

can be simulated by using its relationship to the exponential distribution.

The Poisson distribution represents the number of events that occur in a unit time interval when the time between events has an exponential distribution with parameter λ. Thus, to simulate a value of a Poisson random variable we sum simulated values from the related exponential distribution until that sum is > 1. In this case the value of the Poisson random variable is one less than the number of terms in the sum.

Thus, if $E_1, E_2, E_3 \ldots$ are values simulated from an exponential distribution with parameter λ, then if

$$S(k) = \sum_{i=1}^{k} E_i$$

the first k for which $S(k+1) > 1$ has a Poisson distribution.

We can program this as follows:

```
s := 0; k := -1;
REPEAT
   e := -(1 / lambda) * log(random);
   k := k + 1;
   s := s + e;
UNTIL s > 1;
```

If we examine this in detail, we see that we need to calculate the logarithm of a value each time we go round the loop. This can become expensive. However, with a little thought before we write the program we can avoid this.

We choose k to be the first k such that $S(k+1) > 1$

That is $-(1 / \lambda) \sum \log_e U_i > 1$

Thus $\sum \log_e U_i < -\lambda$

Therefore $\log_e(U_1 U_2 \ldots U_{k+1}) < -\lambda$

That is $U_1 U_2 \ldots U_{k+1} < \exp(-\lambda)$

Thus we require the first k for which the product of $k+1$ uniform values becomes less than $\exp(-\lambda)$, which can be programmed as

```
e := exp(-lambda);
product := 1;
k := -1;
REPEAT
  k := k + 1;
  product := product * random;
UNTIL product < e;
```

From this we see that the calculation of a logarithm each time we pass through the loop is replaced by the calculation of a single exponential, hence the program will perform the operation much more quickly.

15.4 Exercises

15.1. Suppose a coin is tossed 10 times. How many combinations of heads and tails would give totals of 0,1,2 ... 10 heads respectively.

Write a program to simulate the coin tossing situation described above. Construct a frequency distribution for the number of heads observed in ten tosses. How closely do these match your theoretical values after 10, 100, 1000 and 10 000 experiments?

15.2. Generate 1000 random points in a unit square and, by observing how many fall between the curve $y = x \ln(1+x)$ and the x-axis, estimate

$$\int_0^1 x \ln(1+x) \, dx$$

Using a further 1000 points, refine your estimate and comment on the accuracy of your method.

15.3. Write a program that estimates π by simulating the points in a unit square and estimating the area of a circle inscribed in that square by

$$\frac{\text{Number of points inside the circle}}{\text{Total number of points simulated}}$$

15.4. By simulating a sequence of values that are uniformly spread over the interval $[-0.5, 0.5]$, simulate the sum of two and three such values. Evaluate your results in terms of the numerical problems described in chapter 5.

15.5. Write a function to generate random integer values.

15.6. Estimate, using a random integer generator, the probability that two random integers have a common divisor greater than one.

15.7. Write a program to give school children practice in the basic skills of arithmetic. Your program should be able to pose a range of questions based on simple integers. The problems should be randomised by using a random number generator to generate the integers. It should also be able to keep score and report at the end on the child's performance. Allow up to three attempts on each question and grade the score from each question accordingly. (Avoid subtractions that result in negative answers and divisions that result in fractions.) For problems involving integers containing two or more digits, allow the program to present the problem as either 27 + 37, for example, or in the tens and units style of problem.

Modify your program to allow for a range of skills. For example, lower-level examples should avoid things such as 'carry overs' and borrowing.

15.8. A random walk is described by the values of the sequence

$$S_n = X_1 + X_2 + \ldots + X_n$$

where $X_k = \begin{cases} 1 \text{ with probability } p \\ -1 \text{ with probability } q = 1 - p \end{cases}$

Write a program to simulate a random walk and in particular to display the number of steps between each return to the origin when $p = 1/2$.

15.9. Two gamblers A and B play a sequence of games of chance in which the probability of A winning any game is p and the probability of B winning is $q = 1 - p$. After each game the loser pays the winner £1. If A starts ·with £a and B starts with £b, write a program that simulates the results of a sequence of games. For various values of p and q simulate the distribution of the number steps until the game ends for various values of p and q.

16 Records, Complex Numbers and the Fourier Transform

16.1 Introduction

We have already met arrays as an example of a structured type of linked elements of the same type which can be referenced by a single name, the array name. Pascal, however, allows us to define structures whose linked elements are of different types. Thus, for example, if we were considering a queue which contained high and low priority arrivals, then as well as storing the arrival time in the queue it would also be useful to store the priority, perhaps as a character, H or L say, in such a way that both pieces of information are firmly linked together. Such structures are called RECORDS. When we come to consider queues in a later chapter, this ability to define a structure of linked elements of mixed types is very useful.

As well as applications to queues and related problems that we will consider later, records have far-reaching implications for the construction and management of databases, and help to make Pascal a rich and widely applicable language. They also allow us to define types to represent mathematical quantities such as complex numbers and rational numbers which are not pre-defined in Pascal.

A record consists of a set of fields, which may be of different types, all relating to a single unit. For example, if we wanted to create a database of meteorological data we might require fields corresponding to hours of sunshine, temperature, rainfall — each of which is represented by a real quantity. We might also include fields relating to a general description of the day's weather, wind direction etc. which are best stored as characters or user-defined types. Each record would relate to a single day.

A record is a type definition, the general format of which is

```
TYPE   recordname = RECORD
                        datafields : datatype;
                    END;
```

Note: there may be more than one datafield and a mixture of data types.

Thus, for example, we may set up a type for our meteorological record as follows:

```
TYPE weathertype = (cloudy,sunny,rain,showers,
                                          snow,fog);
     winddirection = (N,NE,E,SE,S,SW,W,NW);
     dailyweather = RECORD
                    sunshinehours,rainfall : REAL;
                    mintemperature,
                            maxtemperature : REAL;
                    weatheram,
                          weatherpm : weathertype;
                    wind : winddirection
                    END;
VAR day : dailyweather;
```

An item in a record can be referenced by the identifier of type *recordname*, followed by a dot then the field name. Thus the field *maxtemperature* of *day* is referenced by

```
day.maxtemperature
```

and as such can be used in the same way as any other variable in a Pascal program. The use of the prefix *day* links it to the other components of that record. Thus, reference to a field of a record consists of a double-barrelled name and we cannot usually refer simply to the field name. Thus in the above example

```
maxtemperature := 23;
```

will usually not be allowed. Now if we wish to perform a whole series of operations on different fields of the same record, then repeated use of double-barrelled names will become rather tiresome. Fortunately, there is one command construction in Pascal which allows us to refer to a series of field names without specifying the record prefix each time. This is the WITH construction which has the form

```
     WITH recordname DO
       BEGIN
           set of statements in which we perform operations
           on some or all of the fields of the recordname
       END;
```

In the set of statements inside this construction, if a field is referenced by name only it is assumed to be the field for the record named at the beginning of the construction. If a field of any other record is referenced we must use its full name, complete with the record name prefix.

Thus, for example, if we wanted to read information into our daily record we could use

```
WITH day DO
  BEGIN
    READ(sunshine,rainfall,maxtemperature,
                          mintemperature)
  END;
```

rather than

```
READ(day.sunshine,day.rainfall,day.maxtemperature,
                          day.mintemperature);
```

Note: as with similar Pascal constructions, if there is only one statement involved then the BEGIN and END can be omitted.

We have already seen that items within a record are referenced by both recordname and fieldname. For obvious reasons we cannot assign values to the recordname. However, if two variables are of the same record type, we can assign one to another provided that the record on the right-hand side has values assigned to each of its component fields.

A similar restriction applies when we consider using records in procedures and functions. Here, records can be used as value and variable parameters in both; however, since a function is used to return a single value, it cannot be of type RECORD.

16.2 Complex Arithmetic

Pascal, unlike some other languages, does not have a pre-defined type to represent complex numbers. However, this is not a problem since we can define our own types. Two possibilities are:

(1) to use a two-dimensional array to represent the real and imaginary parts of a complex number, or
(2) to use a record consisting of two fields.

The latter turns out to be more convenient, particularly when we wish to consider a set of complex numbers. We can easily work with an array of records of the complex type but working with an array of two-dimensional arrays — that is, a matrix of two columns — can lead to confusion if we are not very careful. The WITH construction also helps in making the records solution more attractive. However, we cannot use operators like * and / with simulated complex numbers based on records, which makes the construction of results a bit verbose.

We can define a complex type as follows:

```
TYPE complex = RECORD
                 re, im : REAL;
               END
```

We can now consider the construction of a set of procedures and functions to perform the basic operations of complex arithmetic.

Complex addition consists of adding the real parts together and the imaginary parts together. The procedure will require two value parameters of type complex containing the original values and a variable parameter also of type complex to receive the result.

```
PROCEDURE complexaddition(a, b : complex;
                               VAR ans : complex);
BEGIN
   ans.re := a.re + b.re;
   ans.im := a.im + b.im
END;
```

Subtraction only requires a change of sign which could be handled by a fourth parameter of integer type which takes value 1 or -1 and multiplies the b value at each of the two steps.

Multiplication requires a similar set-up to the addition procedure.

```
PROCEDURE multiplycomplex (a,b : complex;
                               VAR ans : complex);
BEGIN
   ans.re := a.re * b.re - a.im * b.im;
   ans.im := a.re * b.im + a.im * b.re
END;
```

The modulus of a complex number is a real value; hence we can use a function in this case.

```
FUNCTION modulus(a:complex) : REAL;
BEGIN
   modulus := SQRT(a.re * a.re + a.im * a.im);
END;
```

16.3 Discrete Fourier Transform

Suppose we have observations $x_0, x_1 \ldots x_{n-1}$ taken at n equally spaced time points $t = 0, 1 \ldots n-1$, and we want to decide whether there is any periodicity or cycle in the data.

A suitable model for this is the discrete Fourier series

$$x_t = \sum_{j=0}^{n-1} C_j \exp(2\pi i jt/n) \qquad \text{where } i = \sqrt{(-1)} \qquad (16.1)$$

which expresses the data as a sum of discrete sine waves.

The C_j values are complex numbers called the Fourier coefficients of x_t.

Since $\exp(i\theta) = \cos\theta + i\sin\theta$, this can also be written as

$$x_t = \sum_{j=0}^{n-1} C_j \left(\cos(2\pi j t/n) + i\sin(2\pi j t/n)\right)$$

Thus, x_t is represented as the complex sum of periodic functions. The frequency $2\pi/n$ is called the fundamental frequency and represents a single cycle which occurs in the time span of the data. The frequencies $2\pi j/n$ $(j = 2 \ldots n-1)$ are called the harmonics of the fundamental frequency and represent cycles that occur j times within the duration of the data. C_0 is a constant while $C_1 \ldots C_{n-1}$ represent the weight that each harmonic contributes to the model. If an unknown cycle is present in the data, then the C_j values for frequencies $2\pi j/n$ in the neighbourhood of the unknown frequency will be larger than the others. Hence, by examining C_j or C_j^2, which is the amplitude of the fitted waves, we can identify possible cycles that are present in the data.

In the next section we consider the problem of calculating the C_j values and in the final section we illustrate its use.

16.3.1 Estimating the Fourier coefficients

Given the observed values, we want to estimate the Fourier coefficients C_j. If we multiply equation (16.1) by $\exp(-2\pi i k t/n)$ and sum over $t = 0 \ldots n-1$, then we get

$$\sum_{t=0}^{n-1} x_t \exp(-2\pi i k t/n) = \sum_{t=0}^{n-1} \sum_{j=0}^{n-1} C_j \exp(2\pi i(j-k)t/n) \qquad (16.2)$$

Consider the sum

$$\sum_{t=0}^{n-1} \exp(2\pi i(j-k)t/n)$$

If $j = k$ then the sum has the value n since the exponential has value 1 for all t. If $j \neq k$ then the sum is a finite geometric series with common 'ratio' $\exp(2\pi i(j-k)/n)$ hence the sum equals

$$\frac{1 - \exp(2\pi i(j-k))}{1 - \exp(2\pi i(j-k)/n)}$$

the numerator of which is zero. Hence equation (16.2) reduces to

$$\sum_{t=0}^{n-1} x_t \exp(-2\pi i j t/n) = C_j$$

which is called the discrete transform of the series of observations $\{x_t\}$.

If the series of observations $\{x_t\}$ is real, then C_0 and $C_{n/2}$ (if n is even) are both real

$$C_0 = (1/n) \sum_{t=0}^{n-1} x_t = \bar{x}$$

while $C_{n/2} = (1/n) \sum x_t \exp(-\pi i t) = (1/n) \sum x_t (-1)^t$

While for $j = 1 \ldots [(n-1)/2]$

$$C_{n-j} = (1/n) \sum_{t=0}^{n-1} x_t \exp(-2\pi i(n-j)t/n)$$

$$= (1/n) \sum_{t=0}^{n-1} x_t \exp(-2\pi i t) \exp(2\pi i j t/n)$$

$$= (1/n) \sum_{t=0}^{n-1} x_t \exp(2\pi i j t/n) = C_j^*$$

Thus C_{n-j} is the complex conjugate of C_j. Hence we need only calculate C_j for $j = 0 \ldots [(n-1)/2]$ in the real case which considerably reduces the amount of computation. Hence we should exploit these features when writing a program to perform the calculations.

Before writing the program, there is another feature of the problem that can save us time in the computation. The term $\exp(-2\pi j t/n)$ can be represented as a complex number whose real part is $\cos(2\pi j t/n)$ and imaginary part is $-\sin(2\pi j t/n)$. Now, since cosine and sine waves are periodic functions, we only need to consider the frequencies $2\pi j/n$ for $j = 0$ to $n-1$ — that is, the frequencies in $[0, 2\pi]$. Any multiple of j which is greater than n can be reduced modulo n, so that it lies in the range $[0, 2\pi]$. Thus, rather than recalculate the values of the complex number each time it is required, we can produce a more efficient program by storing the values $\exp(-i2\pi j/n)$ in an array of complex records for $j = 0 \ldots n-1$ before starting to calculate the Fourier coefficients. Then each time a value of $\exp(-i2\pi j t/n)$ is required, we simply use location $(jt \text{ MOD } n)$ in our array.

If the time series x_t represents the real part of a complex number, then the calculation on the right-hand side is simply the sum of products of complex numbers. Hence, in terms of writing a program we can simplify this into the following steps, using records of complex type to store the data values, the complex values of $\exp(-i2\pi j/n)$ for $j = 1$ to $n/2$ and the resulting Fourier transform.

> Store the values of $\exp(-i2\pi j/n)$ for $j = 0$ to n
>   in an array of type complex
> For each of the frequencies
>   set the Fourier transform record equal to zero
>   FOR each time observation
>   Calculate the complex product of x_t and $\exp(-2\pi j t/n)$
>   add the result to the Fourier transform
>   Multiply the Fourier transform by the scalar $1/n$

This gives us the basis for the following procedure where *data* is a declared type of arrays of complex records defined by

```
TYPE   data = ARRAY[0..maxnoofobservations] OF complex;
```

The type complex is defined in section 16.2.

The parameters *x, n*, and *fouriertransform* are used for the original data, the number of observations and the calculated Fourier transform respectively.

```
PROCEDURE dft (x : data; n : INTEGER; VAR a : data);
(* procedure to calculate the fourier transform *)

VAR w : ARRAY[0..maxnoofobservations] OF complex;
    t,j : INTEGER;
    ang : REAL;
    temp : complex;

BEGIN

   (* calculate 2 *  π/ n *)
   ang:=arctan(1)*8/n;

   (* calculate the values of exp( -i2π j/n)   *)
   FOR j := 0 TO n-1 DO
     WITH w[j] DO
       BEGIN
           re := COS(ang * j);
           im := -SIN(ang * j);
       END;

   (* calculate  the   fourier  transform for each of the
      frequencies *)
   FOR j :=0 to TRUNC(n - 1) / 2) DO
     BEGIN

   (* set the fourier transform to zero *)
       WITH a[j] DO
         BEGIN
             re := 0.0;
             im := 0.0;
         END;

    (* sum the fourier transform *)
       FOR t := 0 to n-1 DO
         BEGIN
           multiplycomplex(x[t],w[t*j MOD n],temp);
           complexaddition(a[j],temp,a[j]);
         END;

   (* perform the scalar multiplication *)
       scalarmult(a[j],1/n,a[j]);
     END;

END;
```

The procedure uses three procedures which must be declared above it in the program. These are the procedures *complexaddition* and *multiplycomplex* as described in section 16.2 which calculate the sum and product of complex numbers, and the procedure *scalarmult* given below which multiplies a complex number by a scalar.

```
PROCEDURE scalarmult(a:complex; scalar:REAL; VAR b:complex);
BEGIN
    b.re := scalar * a.re;
    b.im := scalar * a.im;
END;
```

16.3.2　Applications of the Fourier transform

A simple cycle is described by a cosine wave of the form

$$x_t = R \cos(2\pi ft + \phi)$$

where R is the amplitude of the wave, f is the frequency and ϕ is its phase. If our data consists simply of observations that lie on this wave for various values of t and the frequency f is one of the harmonics of the fundamental frequency used in our Fourier series model, then only the C_j values at that frequency will be non-zero. If the phase ϕ is zero, then only the real part is non-zero. Thus, table 16.1 shows the estimated Fourier coefficients when $f = 0.25$. The first set

Table 16.1　Estimated Fourier coefficients from three known models

Harmonic	Frequency = 0.25 phase = 0		Frequency = 0.25 phase = 2		Frequency = 0.17 phase = 2	
0	−0.000	0.000	−0.000	0.000	−0.058	0.000
1	−0.000	0.000	−0.000	−0.000	−0.060	−0.003
2	−0.000	0.000	−0.000	−0.000	−0.066	−0.006
3	−0.000	−0.000	−0.000	−0.000	−0.082	−0.012
4	−0.000	0.000	−0.000	−0.000	−0.123	−0.023
5	−0.000	0.000	−0.000	−0.000	−0.361	−0.082
6	−0.000	−0.000	0.000	−0.000	0.256	0.068
7	−0.000	−0.000	0.000	−0.000	0.084	0.025
8	0.500	−0.000	−0.208	0.455	0.047	0.015
9	0.000	−0.000	−0.000	0.000	0.031	0.011
10	0.000	0.000	−0.000	0.000	0.023	0.008
11	0.000	0.000	−0.000	0.000	0.018	0.006
12	0.000	0.000	−0.000	0.000	0.015	0.004
13	0.000	−0.000	−0.000	0.000	0.013	0.003
14	0.000	0.000	−0.000	0.000	0.011	0.002
15	0.000	0.000	−0.000	0.000	0.011	0.001

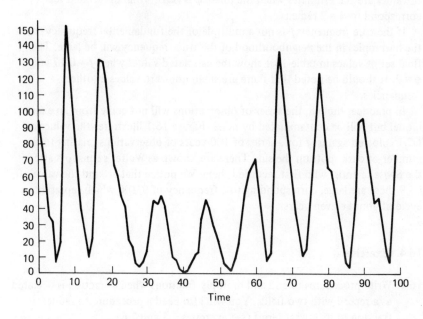

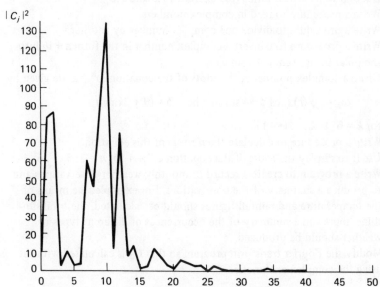

Figure 16.1   Wolfer's sunspots for 1770 to 1869, and the amplitude $|C_j|^2$ of their discrete Fourier transform

of values are the estimates when the phase ϕ is zero, while the second set correspond to $\phi = 2$ radians.

If the true frequency f is not a multiple of the fundamental frequency, then the harmonics in the neighbourhood of the true frequency will be large. The final set of values in table 16.1 show the estimated values when $f = 0.17$ and $\phi = 2$. It should be noted that there are small non-zero values at other frequencies.

In practice, though, the series of observations will not come from an exact model but will be contaminated by noise. Figure 16.1 illustrates the values of $|C_j|^2$ plotted against j for a series of 100 years of observations relating to the number of dark spots on the sun. These are known as Wolfer's sunspots after the Swiss scientist who first recorded them. We notice that around the value of $j = 9$ these are large, corresponding to a frequency of 9/100 which represents a cycle of about eleven years.

16.4 Exercises

16.1. Write procedures to add and multiply fractions when a fraction is declared as a record with two fields. You will also need a procedure to reduce a fraction to its lowest terms (see exercises **7.5** and **7.6**).

16.2. Write a procedure to read in complex numbers.

16.3. Write a procedure to divide one complex number by another.

16.4. Write a procedure to convert a complex number in the form $a + ib$ into the polar form $r\,(\cos\theta + i\sin\theta)$.

16.5. Given a complex number z, the roots of the equation $\omega^n = z$ are given by

$$\omega_k = \sqrt[n]{(r)}\,(\cos\phi + i\sin\phi) \text{ where } \phi = (\theta + 2k\pi)/n$$

for $k = 0, 1, 2 \ldots (n-1)$.

Write a procedure to calculate the n roots of this equation.

Use it to display the roots of the equation $\omega^n = 1$ for $n = 2, 3 \ldots 6$.

16.6. Write a program to create a record of monthly weather observations and to produce a summary of the observations. For example, the means of the temperature and rainfall figures should be calculated, the total sunshine hours and a summary of the occurrences of different types of weather should be produced.

16.7. Modify the Fourier transform program so that C_0 is calculated without using the complex arithmetic.

17 Pointers and Polynomials

17.1 Introduction

The Pascal identifier types that we have met so far are all examples of fixed structures in the sense that, when they are declared at the head of a program segment, they allocate a fixed amount of space into which we can store values. We have met simple types such as REAL and INTEGER, and structured types such as ARRAYs and RECORDs. Each of these has a specific role to play and contributes to the richness and versatility of the Pascal language. However once a program is compiled and running, if we try to access any element outside the declared bounds, an error will be caused. There is nothing we can do but alter the program. Thus the types already discussed cannot easily cope with situations in which we are not sure beforehand how large our structures are likely to be.

The order of the elements in a structure such as an array is also fixed and this makes it difficult to insert or remove terms from the main body of the array. For example, if we delete a term from an array we need either to shift up the rest of the array to close the gap or devise some method of keeping track of empty positions.

The Pascal language, however, includes further types called pointer types which allow us to handle dynamic structures. These are structures whose size can be altered while the program is running, not fixed when it is compiled. Thus, they can adapt to situations in which the size of an entity is not known before starting to run the program. We can dynamically expand or contract the structure, add and remove bits of it, and generally rearrange it in quite complicated ways.

In what ways are such facilities going to be useful to us when writing programs?

The more obvious applications come from areas away from the area of numerical mathematics and include construction and management of databases for use in, say, stock control, travel agencies or libraries etc. There are, however, some examples in mathematics which become much easier to handle and more flexible using this facility. For example, a queue can be thought of as a dynamic structure which items either leave or join. Tree and graph manipulation is another area where the relationship between components is not known at the outset. There are also applications involving matrices which contain many zero terms, called sparse matrices. Using a dynamic structure rather than an array we need only save information about the non-zero elements, thereby saving space, which may be crucial in cases when the matrices are large. It also has the added advantage that we are forced to think in terms of only the non-zero elements, thereby avoiding any unnecessary operations on the zero ones.

To illustrate the use of pointers we will write a program to perform algebraic manipulations on polynomials. From the above preamble, we see that using a dynamic structure to represent each polynomial, we shall not need to allow space for terms with zero coefficients nor need we decide before running the program the maximum order of polynomial that we are going to consider.

17.2 Pointer Types

A variable of pointer type differs from all the other types already described since, instead of being used to hold a *value* of a variable, it is used to hold an *address* in the computer's memory at which the value is actually stored. Thus it points to values rather than contains values. The advantage of this is that by creating, destroying and changing this address we can manipulate the dynamic structures alluded to above.

Typically, pointers point to RECORD variables in which at least one component is a pointer capable of pointing to another such record: it is only at this degree of complication that pointers are of real use. However, to simplify our introduction to the definition and use of pointers we will describe pointers to a single value and in particular pointers to INTEGER variables.

A pointer type (in this case one capable of pointing to INTEGER variables) is defined by

```
TYPE integerptr = ^INTEGER;
```

The type to the right of the circumflex ^ (an up arrow ↑ could be used instead), the domain type, is the type of variable pointed at by the pointer variable whose type is being defined. Variables *p*, *q* of type *integerptr* can now be declared by

```
VAR p,q : integerptr;
```

These variables are capable of holding addresses at which INTEGER values can be stored, but like any other declared variable they are at this stage unassigned. An error would result if they were used in this state.

The only way of creating a new storage location is by using the pre-defined procedure NEW. Thus

```
NEW(p);
```

would create a new address which would be assigned to the *integerptr p*.

Once this is created, we could allocate the same address to *q* by the assignment

```
q := p;
```

The only other way of assigning a value to a pointer variable is to assign the value NIL, for example

```
q := NIL;
```

The value NIL is not an address and a pointer assigned this value does not point to a variable. However, the pointer *is* assigned and we can make use of this in constructions such as

```
IF q = NIL THEN ....
```

without incurring an error.

Once our pointers have been declared and assigned addresses we may assign and manipulate the values to which they point. The constructs $p\hat{}$ and $q\hat{}$ refer to the values pointed to by our (INTEGER) pointers p and q (note that the circumflex now follows the name).

Thus

```
p^ := 2;   q^ := 3;  WRITELN(p^,q^,p^*q^);
```

would result in the the values 2, 3 and 6 being displayed.

Note that operations like this may *not* be performed on the pointers p and q themselves. In particular, Pascal takes the view that is is none of our business what address it has allocated and does not allow WRITELN(p).

The only other operation that we are allowed to do with pointers other than those mentioned above is to remove or de-allocate an address. The procedure DISPOSE(p) de-allocates the locations pointed to by p, making them available for re-use. Naturally, it should only be used when the information kept there is no longer required. Conversely, one should get into the habit of using DISPOSE where appropriate, else large programs, run using a version of Pascal which does not automatically re-allocate memory not currently being used, may quite unnecessarily run out of memory.

We will not use DISPOSE in our subsequent discussions since it does not appear to be consistently implemented in various versions of Pascal, but good book-keeping habits should be developed as your programming skills develop.

17.3 Linked Structures

The real power and flexibility of pointer types can be seen when we consider a pointer of RECORD type which contains at least one field which is a pointer of the same type. By doing this we are able to set up linked lists, where the field of pointer type is used to point to the next record in the list. We begin by considering the definition of such structures before discussing their uses.

Consider the following example.

A polynomial can be thought of as a set of terms each described by two values: a real coefficient and an integer exponent or power. If the coefficient is zero, then that particular power does not figure in the polynomial and it is unnecessary to store such information. Thus, one way of thinking of a polynomial in Pascal is as a linked list of records where each record consists of three fields: a real coefficient, an integer exponent, and a pointer to the next term in the polynomial. If there are no more terms in the polynomial then the pointer field is assigned the value NIL, indicating that we are at the end of the chain.

We could therefore define a *polynomial* type as follows

```
TYPE polynomial = ^term;
     term = RECORD
                 coefficient:REAL;
                 exponent:INTEGER;
                 nextterm:polynomial
            END;
```

It should be noted that we have used a type, *term*, before we have defined it which contravenes the general rule that a type may not be used before it is defined. However, in this case reversing the order would mean that the type *polynomial* is similarly used. To resolve this impasse, the following exception to convention is allowed.

> "Pointer type definitions may precede the definitions of their reference type. However the reverse is not true — a structure may not contain a pointer type that has not been defined"

Thus a pointer type must be defined before its type if the type is a user-defined one. Hence the order used in our example is allowed but we could not define *term* before *polynomial*.

The advantage in using pointers of RECORD type, where the record contains a pointer of the same type (as one of its fields), is that it allows us to construct a set of records which are tied together through the pointer fields. Such an arrangement is called a linked structure. Our example is one with a single link which allows us to move in one direction, down the chain. We can, however, have two fields of pointer type allowing us to move in both directions through the list. Extensions of this idea allow us to construct very complicated structures which cannot easily be implemented in some other programming languages.

In general, the definition of such a structure is summarised as follows:

```
TYPE pointername = ^pointertype;
     pointertype = RECORD
                     Datal : fieldtypel;
                     .....
                     Datan : fieldtypen;
                     pointerl,....pointerk :
                                          pointername
                 END
```

The data field names are used to store the values associated with each item in the list. Their field types can be any of the types that we have previously met, such as INTEGER, REAL, ARRAYs etc. The pointer field names are used to point to other items in the structure.

Structures such as our polynomial example can be thought of as a list which can be extended, shortened and have items added or removed from it. In addition to what has already been described, we need a pointer *p* of type *polynomial* to point to the first term of our polynomial. This is illustrated in figure 17.1.

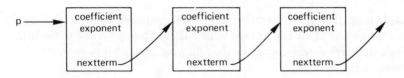

Figure 17.1   Illustration on the links in a list

The list can be created by allocating addresses to *p, p^.nextterm, p^.next-term^.nextterm*, etc., using the procedure NEW. The list is terminated by assigning the value NIL to the last pointer. For example

```
p^.nextterm^.nexterm^.nextterm := NIL;
```

It can be seen that this leads very easily to cumbersome variable names. However, this is overcome by using auxiliary pointers which correspond to the part of the list we are currently working on.

To manipulate our list representing a polynomial we shall make use of an auxiliary pointer of type *polynomial*, called *currentterm*. This will contain the address of the term we are currently working on, as shown in figure 17.2.

We now consider how to perform certain elementary operations on our list.

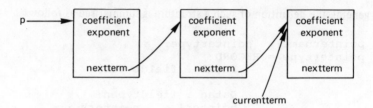

Figure 17.2   Illustration of use of an auxiliary pointer

17.3.1 *Assignment of values to the list*

Any assignments to the locations pointed to by *currentterm* are assignments to the current term in our polynomial.

Thus, for example, the statements

```
currentterm^.coefficient := 3.2;
currentterm^.exponent := 2;
currentterm^.nextterm := NIL;
```

would mean that the last term in our list now contained the values 3.2 and 2, and the pointer at the end had the value NIL, indicating the end of the list. Thus, the three assignments make the following alterations to our list as shown in figure 17.3.

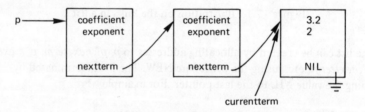

Figure 17.3   The result of assigning values to currentterm

17.3.2 *Extension of the list*

We can easily extend the list by creating an address for the pointer field in the record *currentterm* and then equating this address with *currentterm*. Thus the two statements

```
NEW(currentterm^.nextterm);
currentterm := currentterm^.nextterm;
```

(a) allocate an address to the last link in the chain pointing to another record of type *term*

(b) then assign this last address to *currentterm* so that our auxiliary pointer *currentterm* still points to the last record in the list.

This is illustrated in figure 17.4.

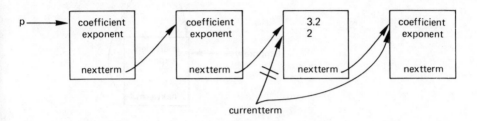

Figure 17.4   The result of appending a new term to the list

As a result of the second statement, any assignments subsequently made to the locations of *currentterm* are placed in the next position in the list and do not overwrite the previous terms.

17.3.3 Insertion of new terms in the list

We can insert a new term in the list, after the current term, by using a second auxiliary pointer called *newterm*. This is required so that we can create the new address before linking it into the chain by redirecting two of the pointers. Thus

NEW(newterm);   Creates a new address associated with *newterm*

newterm^.nextterm := currentterm^.nextterm;

Assigns the address at the end of *currentterm* in *currentterm^.nextterm* to the pointer in the record pointed to by *newterm* so that they both point to the same address, which is the next term in the chain

currentterm^.nextterm := newterm;

Ensures that the exit pointer from the record *currentterm* points to newterm, thus making it the next link in the chain. This breaks one of the two links to the next term

This is shown in figure 17.5.

Note: if these last two statements were in the reverse order, we would lose the link to the rest of the chain since the link from *newterm* would point back to itself. A useful rule in such a situation is only to redirect a pointer which is

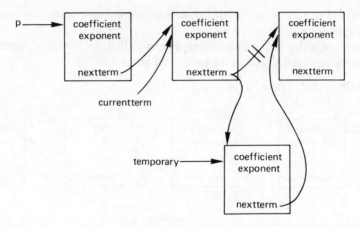

Figure 17.5   Insertion of a new term after the current term

part of the chain when there is a second pointer available to maintain that link. Thus we ensure the link to the remainder of the chain is secure before linking to the head of the term we are including.

To place a new record before the current one, we need to create an address for the new record, move the contents of the current record into the new record, place the new data in the current record and finally insert the new record after the current one. This is achieved by the following steps and is illustrated in figure 17.6.

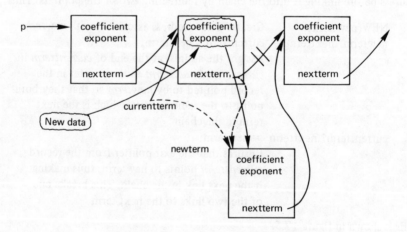

Figure 17.6   Insertion of a record before currentterm

Create a new address   that is, NEW(newterm);
Place the data in the current record in *newterm*
       for example, newterm^ := currentterm^;
Assign new data values to *currentterm*
       for example, currentterm^exponent := newexponent;
Assign the exit from *currentterm* to the exit from *newterm*
       that is, newterm^.nextterm := currentterm^nextterm;
Assign the address of *newterm* to the exit from *currentterm*
       that is, currentterm^.nextterm := newterm;
Move the pointer *currentterm* forward by assigning it the address of *newterm*
       that is currentterm := newterm;

This last step ensures that the pointer *currentterm* is pointing to the same information as it did at the outset.

17.4 Pointers and Procedures

As before, it would be useful to write and use procedures and functions to perform specific operations. However, we need to consider the effect of using pointers as parameters to procedures and functions. There is a slight complication here since associated with each pointer is the address to which it points and the contents of that location.

If the pointer is passed as a parameter, then it is the address that is altered according to the rules applying to parameters. The location, though, is treated as a global variable and hence any change to the contents within a procedure is permanent even if the pointer is passed as a value parameter.

Hence the use of a pointer as a value parameter only partially inhibits the reconstructing ability of the procedure.

Once we are aware of the above difference in operation, we can proceed to structure our problems in the usual way. Thus, if we have a polynomial p into which we want to insert a new term after the current one then, based on the discussion of section 17.3.3 above, we can use procedure *insert* given below.

```
PROCEDURE insert(currentterm : polynomial);
VAR newterm : polynomial;
BEGIN
  NEW(newterm);
  newterm^.nextterm := currentterm^.nextterm;
  currentterm^.nextterm := newterm;
END;
```

17.5 Polynomial Arithmetic

We have seen in chapter 11 how we can handle calculations on polynomials using ARRAYs. However, we needed to declare an array such that its dimensions covered the maximum order of polynomial that we were likely to require. Further, any terms with zero coefficients were also held in the array. If we tackle the problem using a list linked by pointers, then these two problems do not arise. That is, we do not have to decide beforehand the maximum order of polynomial, nor do we store information about terms with zero coefficients. We will consider the problems of addition, and multiplication of polynomials.

However, before performing any operations on polynomials we must begin by writing routines for entering and displaying a polynomial.

17.5.1 Polynomial entry and display

To create a polynomial list we have to read values into the list. A simple way is to read them in decreasing order of the powers or the exponents.

We will use an auxiliary pointer *currentterm*, also of *polynomial* type, to keep track of our position in the list in the way described in the previous section. If *p* is a *polynomial* pointer pointing to the head of the polynomial, then the steps in the process of creating a polynomial list and entering terms into it are:

1. Enter the number of terms in the polynomial.
2. IF the number of terms is zero then assign the value NIL to *p*.
   ELSE
      3. Create an address for the pointer *p*.
      4. Assign the address of *p* to *currentterm* so that the auxiliary pointer points to the head of the list.
      5. Enter each term in the polynomial.
      6. Terminate the list by assigning the value
         NIL to *currentterm^.nextterm*.

Step 2 allows us to handle the entry of zero polynomials.

Step 5 is the main part of the list's construction and can be described as follows

> FOR each of the terms
>> read in the coefficient and exponent of the current term
>> IF we have not reached the last term THEN
>>> create an address for the next record by using the
>>> procedure NEW on *currentterm^.nextterm*
>>>
>>> move the auxiliary pointer forward by assigning
>>> the address of *currentterm^.nextterm* to *currentterm*

Note that this last step is only performed if we need more space to enter further terms into the polynomial. If it is always carried out, then the last address in the list will contain locations for coefficients and exponents which will not have values assigned to them. This is likely to cause runtime errors at a later stage, unless we are very careful, by attempting to use identifiers which have not been defined.

The above can be placed in a procedure *readpolynomial* which requires a single variable parameter *p* of type *polynomial* in which the first address of the polynomial being entered is stored. We also require three local variables, *count* and *numberofterms* of type INTEGER and *currentterm*, a pointer of type *polynomial*, to act as the auxiliary pointer. The result is a procedure such as

```
PROCEDURE readpolynomial(VAR p : polynomial);
VAR
   numberofterms : 0..maxint;
   count : INTEGER;
   currentterm : polynomial;

BEGIN
(* 1. enter the number of terms in the polynomial.   *)
   WRITELN('how many terms are in the polynomial');
   READLN(numberofterms);

   IF numberofterms = 0 THEN  p := NIL
   ELSE
      BEGIN
         WRITELN('now enter the set of coefficients',
                 ' followed by exponents',
                 ' in decreasing order of exponent');
(* 3. assign an address to the  pointer  p, the head of
      the polynomial *)
         NEW(p);
(* 4. assign the  address  of  p to currentterm so that
      the  auxiliary  pointer points to the head of the
      list. *)
         currentterm := p;

(* 5.    Enter each term in the polynomial
         FOR count := 1 TO numberofterms DO
            BEGIN
(*       read in  the  coefficient  and  exponent  of  the
         current term *)
               WITH currentterm^ DO
                  READ(coefficient,exponent);

(*       if the  count  is  still  less  than the number of
         terms create  an  address  for the next record and
         move the auxiliary pointer forward *)
```

```
        IF count < numberofterms THEN
           BEGIN
              NEW(currentterm^.nextterm);
              currentterm := currentterm^.nextterm;
           END;
        END;

(* 6.  Terminate the list. *)
        currentterm^.nextterm := NIL;
     END;
END;
```

It is assumed in subsequent procedures that the terms are entered into the polynomial in decreasing order of the powers. A more robust procedure would check that this was the case or ensure that the terms are placed in the appropriate position in the list. (See exercise **17.2** at the end of the chapter.)

The second procedure that we require is one to display such a polynomial. Again we need an auxiliary pointer to allow us to move down the list which we will declare as a local parameter of type *polynomial*. The procedure will require a single value parameter of type *polynomial* since we make no alterations either to the address or the contents in this procedure. The structure of the procedure is

1. The auxiliary pointer, *currentterm*, is assigned the address of the head of the polynomial so that it is placed at the head of the list.
2. WHILE the auxiliary pointer has not reached the end of the list display each term in turn.

This last step can be broken into

2a.   IF the current coefficient is < 0 THEN Display a $-$ sign
                                    ELSE Display a $+$ sign
          (*note:* we do not have to bother with the case of the
          coefficient having zero value since this case is excluded
          from the list of the polynomial terms).
2b.   IF the current exponent is > 0 THEN display the character x
          IF it is greater than 1 THEN
                display the exponent preceded by the character ^.
2c.   Move the auxiliary pointer on to the next term by assigning
          the address in *currentterm^.nextterm* to *currentterm*.

Translating this into Pascal we have a procedure such as the following:

```
PROCEDURE printpoly(p : polynomial);
VAR currentterm : polynomial;

BEGIN
(* 1.  Place the auxiliary  pointer at the head of the
        list *)
   currentterm := p;
```

```
(*   2.   Display each term in turn.   *)
  WHILE currentterm <> NIL DO
    BEGIN
      WITH currentterm^ DO

(*      2a.   Display the coefficient   and its sign*)
        BEGIN
          IF coefficient < 0 THEN
              WRITE(' - ',ABS(coefficient):4:2)
          ELSE
              WRITE(' + ',ABS(coefficient):4:2);

(*      2b.   If it is not zero display the exponent *)
          IF exponent > 0 THEN WRITE('x');
          IF exponent > 1 THEN WRITE('^',exponent:1)
        END;

(*      2c.   Move the auxiliary pointer  onto the next
              term.  *)
      currentterm := currentterm^.nextterm
    END;
  WRITELN;
END;
```

17.5.2 *Addition of polynomials*

Before considering the main procedures that will add and multiply two polynomials, it is useful to write a small procedure to carry out the operation of
storing the newly created pair of coefficient and exponent values at the current
address. The procedure will also need to move the auxiliary pointer to the next
address, if necessary, and to create a new address to maintain the link to the
next address.

Since each of the main procedures will place the auxiliary pointer at the head
of the polynomial, there will be no need to move it forward when we insert the
first term in the list. If we did so, then the locations at the head of the list
would not be assigned values.

The steps of our procedure, which we will call *attach* because of the operation it performs of attaching the next term to the existing list, are given below.
The variable *existing*, of type *polynomial*, contains the address of the previous
term in the list.

1. IF we are not at the first term THEN
        move the pointer forward by assigning the record's link
        existing^.nextterm to the pointer *existing*.
2. Add 1 to the counter recording the number of terms.
3. Assign the values of coefficient and exponent to the
   appropriate fields in the record.
4. Create an address pointing to the next term in the list, that is
        NEW(existing^.nextterm);

The procedure requires four parameters. Two will be value parameters to receive the values of the current coefficient and exponent, while the other two will be variable parameters. The first of these will be of *polynomial* type to receive and return the address of the existing term, while the other will be of INTEGER type to receive the number of the current term and return the number of the next term.

```
PROCEDURE attach(coefficient : REAL;
                 exponent : INTEGER;
                 VAR existing : Polynomial;
                 VAR termnumber : INTEGER);
BEGIN
  IF termnumber <> 1 THEN
                 existing := existing^.nextterm;
  termnumber := termnumber + 1;
  existing^.coefficient := coefficient;
  existing^.exponent := exponent;
  NEW(existing^.nextterm);
  END;
```

This procedure will be used in both the addition and multiplication procedures once the next term in the polynomial has been evaluated. We begin by considering a procedure to add two polynomials, p and q.

In algebraic terms, given the two polynomials

$$p(x) = a_n x^n + a_{n-1} x^{n-1} + \ldots + a_1 x + a_0$$

and $q(x) = b_r x^r + b_{r-1} x^{r-1} + \ldots + b_1 x + b_0$

We have $p(x) + q(x) = \sum_{j=0}^{k} (a_j + b_j) x^j$ where $k = \max(n, r)$.

Note: $a_j = 0$ if $j > n$ and $b_j = 0$ if $j > r$.

Thus we simply have to consider all the powers in the polynomials, adding the appropriate coefficients.

We will use a single auxiliary pointer to keep track of our position in the list representing the sum of the polynomials. The approach is to move down our two polynomials and form our sum polynomial from exponents that are present in each of them. Thus, if an exponent value is present in both we add the coefficients associated with that exponent, and if the resulting coefficient is not zero we attach the term to our new polynomial and move forward to the next term of both the polynomials. If the exponent value is only present in one of the existing polynomials, we simply add that term to our new polynomial and move to the next term of that polynomial. Once we reach the end of at least one of our two polynomials we must include any remaining terms, from any polynomial whose end has not been reached. The structure can be summarised as follows:

1. Create a new address for the head of the sum polynomial.
2. Assign this address to our auxiliary pointer.
3. Set the number of terms to 1.
4. WHILE we have not reached the end of either of the two polynomials that we are adding, we calculate the next term in the polynomial.
5. Once we have reached the end of at least one of our two polynomials, we must include any remaining terms from that polynomial, if any, whose end we have not reached.
6. Set the exit address of the auxiliary pointer to the value NIL to indicate the end of the list.

The first three steps are straightforward. Step 4 consists of finding the next exponent in decreasing order and then either including the single contribution from one of the polynomials or including the term with the sum of the coefficients, provided it is non-zero, if the exponent occurs in both of our polynomials. Thus

> IF the next exponent of polynomial p is less than the next
> > exponent of polynomial q THEN
> > > Attach the next term of q to our new polynomial and move to
> > > > the next term of polynomial q
> ELSE
> > IF the next exponent of polynomial q is less than the next
> > > exponent of polynomial p THEN
> > > > Attach the next term of p to our new polynomial and
> > > > move to the next term of polynomial p
> > ELSE
> > > IF the sum of the two coefficients is non-zero THEN
> > > > include the sum of the coefficients as the next
> > > > term in our new polynomial;
> > > move to the next term of both the p and q polynomials.

The resulting procedure will require three parameters: two value parameters p and q of *polynomial* type which will point to our two existing polynomials and a variable parameter *sum*, also of *polynomial* type, which will point to our new polynomial resulting from adding the other two. We will also need two local parameters, namely *currentterm*, of *polynomial* type for use as the auxiliary pointer, and *termnumber*, of INTEGER type which will be used to keep track of the term on which we are currently working. Our procedure is

```
PROCEDURE polyadd(p,q : polynomial;
                  VAR sum : polynomial);
VAR currentterm : polynomial;termnumber : INTEGER;

BEGIN
(* initialize the pointers and the counter to indicate
   the number of terms *)
```

```
        NEW(sum);
        currentterm := sum;
        termnumber := 1;
(* calculate the next term in the polynomial *)
     WHILE (p <> NIL) AND (q <> NIL) DO
        IF p^.exponent < q^.exponent THEN
           BEGIN
              attach(q^.coefficient,q^.exponent,
                                currentterm,termnumber);
              q := q^.nextterm;
           END
        ELSE
           IF p^.exponent > q^.exponent THEN
              BEGIN
                 attach(p^.coefficient,p^.exponent,
                                currentterm,termnumber);
                 p := p^.nextterm;
              END
           ELSE
              IF p^.coefficient + q^.coefficient <> 0
                 THEN
                    BEGIN
                       attach(p^.coefficient
                            +  q^.coefficient, p^.exponent,
                                     currentterm,termnumber);
                       p := p^.nextterm;
                       q := q^.nextterm;
                    END;

(*  include  any  remaining  terms  from  the  original
    polynomials *)
     WHILE p <> NIL DO
        BEGIN
           attach(p^.coefficient,p^.exponent,
                                  currentterm,termnumber);
           p := p^.nextterm;
        END;
     WHILE q <> NIL DO
        BEGIN
           attach(q^.coefficient,q^.exponent,
                                  currentterm,termnumber);
           q := q^.nextterm;
        END;

(* assign value NIL to the pointer at the list's end *)
     currentterm^.nextterm := NIL;
END;
```

17.5.3 *Multiplication of polynomials*

In algebraic terms the multiplication of two polynomials

$$p(x) = a_n x^n + a_{n-1} x^{n-1} + \ldots + a_1 x + a_0$$

and $q(x) = b_r x^r + b_{r-1} x^{r-1} + \ldots + b_1 x + b_0$

results in the polynomial

$$c(x) = p(x) * q(x)$$

$$= c_{n+r}x^{n+r} + \ldots + c_1 x + c_0$$

where $c_k = a_0 b_k + a_1 b_{k-1} + \ldots + a_k b_0 = \sum_{j=0}^{k} a_j b_{k-j}$

for $k = 0 \ldots n+r$

As before, $a_j = 0$ if $j > n$ and $b_j = 0$ if $j > r$.

Thus the multiplication of two polynomials is complicated by the fact that we need to perform several passes through each of our existing polynomials. Therefore, as well as the auxiliary pointer used to indicate our position in the new polynomial's list, we also need two other auxiliary pointers which will be used to move down each of our existing two polynomials. The main structure of the operation of multiplying two polynomials together can be summarised as

1. Create an address for the new polynomial.
2. Assign this address to the auxiliary pointer *currentterm* which we will use to keep track of our result.
3. Set the *termnumber* to 1.
4. Start at the highest exponent by setting *nextexponent* to the sum of the exponents pointed to by p and q, that is set
    nextexponent := p^.exponent + q^.exponent
5. WHILE the value of the next exponent is not negative, calculate the coefficient of the term associated with that exponent and if it is not zero include the resulting term in our new polynomial.
6. Terminate the new polynomial list by assigning the value NIL to *current^.nextterm*, the exit from the current record.

This gives us the basic structure of the procedure, leaving us to concentrate on step 5, where the main part of the work is performed. This step can be broken down in the following way:

5a. Set the value of the next coefficient to zero.
5b. Move the auxiliary pointer associated with p to the head of p by assigning the address of p to it.
5c. Taking each term in polynomial p in order, search for the term in the polynomial q such that the sum of the exponents equals the value of the exponent on which we are currently working.
    IF such a term is found THEN
            add the product of the coefficients
            to the current value of the next coefficient
5d. IF the resulting value of the next coefficient is non-zero
            THEN include the term in our new polynomial.
5e. Reduce the value of the next exponent by 1.

This leaves us with step 5c to break down. It is best thought of as a repeat loop, the operations being performed until we reach the end of the polynomial *p*. That is, until the auxiliary pointer, *currentp*, associated with *p* has the value NIL. The steps of the repeat loop are

5c. (i) Assign the difference between the value of next exponent and the current value of the *p* exponent to a variable *remexponent*, say.

5c. (ii) IF this difference is not negative THEN

a. set the auxiliary pointer, *currentq*, associated with *q* to the head of *q* by assigning *q*'s address to it.

b. WHILE the current exponent of *q* is greater than *remexponent* AND we have not reached the end of *q* move to the next term of *q*.

c. IF we have reached an exponent of *q* equal to *remexponent* (we have found a term which contributes to our next coefficient) THEN

add the product of the current p and current q coefficients to our value of the next coefficient.

The result of this is to create a procedure as follows:

```
PROCEDURE Polymult(p,q : polynomial;
                   VAR mult : polynomial);
VAR currentterm,currentp,currentq : polynomial;
    currentexponent,remexponent,termnumber : INTEGER;
    currentcoefficient : REAL;

BEGIN
(* if either  of  the  two polynomials is zero then the
   result is zero *)
 IF (p = NIL)  OR (q = NIL) THEN mult := nil
 ELSE

(*  steps 1-4 initialise the pointers  associated  with
the  answer  and  the variables that indicate the number
of term and the current exponent  *)
  BEGIN
    NEW(mult);
    currentterm := mult;
    termnumber := 1;
    currentexponent := p^.exponent + q^.exponent;
(*  step 5 calculate  the  coefficient  of  the  current
exponent *)
    WHILE currentexponent >= 0 DO
      BEGIN
        currentcoefficient := 0;
        currentp := p;
(*  step 5c.   For each term in polynomial *)
        REPEAT
          remexponent := currentexponent -
                              currentp^.exponent;
```

```
(*    step  5c(ii).  find any term in q  which  combines
with  the  current  term  in q  to  contribute  to  the
coefficient of the current exponent *)
          IF remexponent >= 0 THEN
             BEGIN
                currentq := q;
                WHILE (currentq^.exponent > remexponent)
                      AND (currentq^.nextterm <> NIL) DO
                   currentq := currentq^.nextterm;
                IF currentq^.exponent = remexponent THEN
                   currentcoefficient := currentcoefficient
                                   + currentp^.coefficient *
                                         currentq^.coefficient
             END;

          currentp := currentp^.nextterm;
        UNTIL (currentp = NIL);

        IF currentcoefficient <> 0 THEN
           attach(currentcoefficient,nextexponent,
                              currentterm,termnumber);
        currentexponent := currentexponent - 1;
    END;

(*    step  6.  terminate  the  list  of  the  resulting
polynomial *)
    currentterm^.nextterm := NIL;
  END;
END;
```

17.5.4 Example

We conclude this chapter by illustrating the use of these procedures. Suppose we have two polynomials

$$p(x) = 3x^2 + x + 4$$
$$q(x) = 4x^5 + 3x^3 + 2x + 3$$

and we want to form their sum and product. We can do this using the following commands:

```
BEGIN

  (* read and display the two polynomials *)
  readpolynomial(p);
  readpolynomial(q);
  WRITE('p(x) =');
  printpoly(p);
  WRITE('q(x) =');
  printpoly(q);
```

```
(* perform the addition *)
polyadd(p,q,sum);
WRITE('p(x)+q(x) = ');
printpoly(sum);

(* perform the multiplication *)
polymult(p,q,product);
WRITE('p(x)*q(x) = ');
printpoly(product);
END.
```

The result of this is the following output:

```
p(x) = + 3.00x^2 + 1.00x + 4.00
q(x) = + 4.00x^5 + 3.00x^3 + 2.00x + 3.00

p(x)+q(x) =  + 4.00x^5 + 3.00x^3 + 3.00x^2 +
3.00x + 7.00

p(x)*q(x) =  + 12.00x^7 + 4.00x^6 + 25.00x^5 +
3.00x^4 + 18.00x^3 + 11.00x^2 + 11.00x + 12.00
```

17.6 Exercises

17.1. Write a procedure to insert a term either before or after the current term in a list.

17.2. Write a procedure to allow entry of the terms of a polynomial in any order, the entry terminating when a negative exponent is given. The procedure should place each term in its correct position in a list by descending order of exponents so that the resulting polynomial can be used in the other procedures described in this chapter.

17.3. Write a procedure to evaluate a polynomial, which is stored in a list, for a particular value of x.

17.4. Write a program to generate a list of primes, using the sieve of Eratosthenes described in chapter 11. In this case, use a list in which to store the values so that no storage is wasted once the non-primes are eliminated.

17.5. Using the procedure to create a list of primes written as part of the previous exercise, write a program to determine the prime factors of any integer.

18 Simulating Queues

18.1 Introduction

A queueing system consists of items arriving according to some form of arrival distribution to be served by one or more servers. The length of time taken to serve customers is described by one or more given service distributions. The queue discipline determines the way in which customers move through the system, the simplest being a first come first served system. Knowledge of these three things is sufficient to permit modelling of the system mathematically or simulating it on a computer.

We may, for example, be interested in the queue lengths likely to result, the waiting times of the customers or the idle time of the servers. These quantities will usually be represented by a distribution of possible values. For some queueing situations it is possible to calculate the distribution, or perhaps just the mean, in terms of parameters of the arrival and service distributions.

If there is no known mathematical solution then we can often use simulation methods to estimate all these things. Here the techniques discussed in chapter 14 will be useful. In general, the more complex the queue set-up, the more we have to resort to simulation methods — particularly when we wish to consider different queueing rules, varying numbers of servers and different arrival and service distributions.

18.2 Simple Queues

The simplest way to think of a queue is as a list where items are added to the end and removed from the beginning. Thus a simple list that is used to represent a queue requires two pointers of the same record type associated with it. One is used to represent the head of the queue and the other to represent the tail, as shown in figure 18.1. The record will require a single field of pointer type, called *next*, say, which is used to point to the next item in the list.

Thus to store information about a queue, we will need to define the following structure and declare three variables of this type:

237

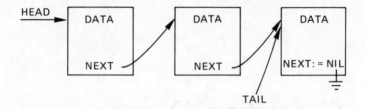

Figure 18.1   Pictorial representation of a queue

```
TYPE queue = ^item;
     item = RECORD
                data field names of data field types;
                next : queue;
            END;
VAR  head,tail,temp : queue;
```

The data field names and their types depend on the information we wish to
record about items in the queue. For example, we will usually wish to record
at least the arrival time.

    Note that a new arrival is inserted at the tail of the queue, while, when a
server becomes free, the next person or item to be served is removed from the
head of the queue.

    If the queue is empty, both the pointers *head* and *tail* have the value NIL.

    If an item is removed from the head of the queue for serving we simply move
the pointer *head* on to the next address using

```
head := head^.next;
```

    However, if this results in the queue being empty — that is, *head* has the
value NIL as a result of this move — we must also set the value of the pointer
tail to NIL. Thus, following the previous statement we also require

```
IF head = NIL THEN tail = NIL;
```

    If a new item is added to the end of the queue, we need to create a new
address using a temporary pointer and ensure that the pointer from that record
has the value NIL.

```
NEW(temp);
temp^.next := NIL;
```

    If the queue is empty when this item is placed in it — that is, if *head* had the
value NIL — then the pointer *head* must be assigned the address of *temp* so that
it points to this new item, otherwise the exit from the existing tail must be

pointed at this new term. That is, we must assign the address of *temp* to *tail^.next*. This is achieved by the statement

```
IF head = NIL THEN head := temp
ELSE  tail^.next := temp;
```

Finally, having ensured that the new item is attached to the end of the existing queue, the tail pointer is then moved on to the new term by the assignment

```
tail := temp;
```

which places it at the bottom of the queue again.

The simulation is performed by handling each event as it happens in time and generating information about the next event of that kind. Thus a simple queue simulation can be summarised by

> at the time of the next event
>> carry out the operations associated with that event and
>> where necessary generate the next event of that kind

This step is repeated until the time of the next event exceeds the maximum time period for the simulation.

For a simple queue with a single server we need to consider events of two different kinds: the arrival of new customers and the completion of service of a customer. In order to find out which kind of event is the next to occur we maintain two variables, *timeofnextarrival* and *timeofserviceend*, so that the next event dealt with corresponds to the smaller of these. Of course, if there are no customers then *timeofserviceend* is not meaningful: to deal with this special case we maintain a **BOOLEAN** variable *serverfree*.

Thus the main part of the simulation is

> IF timeofnextarrival < timeofserviceend OR serverfree THEN
>> Process an arrival
>
> ELSE
>> Process an end of service

which is repeated until the time of the next event is greater than the total time period for the simulation.

When an arrival takes place there are two possibilities. One, we commence its service and calculate the time at which that service ends or, two, if the server is busy, we place the item at the tail of the queue. In both cases we calculate the time when the next arrival takes place. Thus the main steps of processing an arrival are

IF the server is free THEN
        generate the time of end of service
    ELSE
        place the arrival in the queue
        and add one to the queuelength
    generate the time of the next arrival

When an end of service takes place there are again two possibilities to consider. We can either commence the next service if the queue is not empty and generate the time at which that service ends, or we can indicate that the server is free and wait for the next arrival. Thus, processing a service end consists of the following operations:

IF the queue is not empty THEN
        generate the end of the next service
        and remove the first item from the queue
    ELSE
        indicate that the server is free

Finally, we must remember to initialise the variables representing the times of the next arrival and the next service end. The former is assigned the times of the first arrival while the latter has value zero. We also need to create an empty queue and set the queuelength to zero. Values will also be required for the total time and a seed for the randomisation process.

```pascal
PROGRAM simplequeue (INPUT,OUTPUT);
TYPE queue = ^item;
     item = RECORD
                  arrivaltime : REAL;
                  next : queue;
             END;
VAR head,tail,temp : queue;
    timeofnextarrival,timeofserviceend : REAL;
    currenttime,totaltime,waitingtime : REAL;
    queuelength,seed : INTEGER;
    serverfree : BOOLEAN;

FUNCTION random : REAL;
CONST
    a = 293;    c = 1;    m = 65536;
BEGIN
    seed := (a * seed + c) MOD m;
    random := seed / m;
END;
```

```
BEGIN

(* create an empty queue *)
  new(head);
  head := NIL;
  tail := NIL;

(* initialize the various variable used *)
  timeofserviceend := 0;  serverfree := TRUE;
  queuelength := 0;  seed := 374;  totaltime := 5;
  timeofnextarrival := random;

(* perform the simulation *)
  REPEAT
    IF (timeofnextarrival < timeofserviceend) OR
                                    serverfree THEN
  (* process the current arrival if it occurs within
     the total time *)
      BEGIN
        IF serverfree THEN
          (* commence service and calculate end of
             service *)
          BEGIN
            waitingtime := 0;
            timeofserviceend := timeofnextarrival +
                                            random;
            serverfree := FALSE
          END

        ELSE
          (* add arrival to the tail of the queue *)
          BEGIN
            queuelength := queuelength + 1;
            NEW(temp);
            temp^.next := NIL;
            temp^.arrivaltime := timeofnextarrival;
            IF head = NIL THEN head := temp
                ELSE tail^.next := temp;
            tail := temp;
          END;

        (* display the information if necessary *)
        WRITELN(timeofnextarrival:5:2,queuelength:3,
                                    'Arrival':15);
        timeofnextarrival := timeofnextarrival +
                                            random;
      END

    ELSE
```

```
(* process the service end *)
      BEGIN
        WRITELN(timeofserviceend:5:2,queuelength:3,
                'Service end ':15, ' wait = ',
                waitingtime:6:2);
        IF queuelength>0 THEN
      (* remove next item from the head of the queue
         and calculate the end of end of its service *)
          BEGIN
            queuelength := queuelength - 1;
            waitingtime := timeofserviceend -
                                head^.arrivaltime;
            head := head^.next;
            IF head = NIL THEN tail := NIL;
            timeofserviceend := timeofserviceend +
                                                random;
          END
        ELSE
          serverfree := TRUE;
      END;

  UNTIL (timeofnextarrival > totaltime) AND
                    (timeofserviceend > totaltime);
END.
```

18.3 Doubly Linked Lists and Rings

All the examples that we have met so far have been of lists which we have only been able to scan in a single direction. There are many situations where we need to scan a list in both directions.

Suppose we want to simulate a problem which is made up of a whole range of events. Each event is handled as it occurs in the order of time and where possible the next event of that kind is generated. If we have a list of events, then as each new event is generated we place this in its appropriate position in the list according to the time of the event. The current event will be at the head of the list while new events will need to be inserted towards the end of the list. Thus the ability to move in either direction through the list means that we can speed up the process of inserting new events by starting at the end and working backwards.

A list which allows us to move in both directions is called a *doubly linked list*.

A doubly linked list is achieved by having two fields in our record that are pointers of the same type as the record. One of these is used to point forward to the next record while the other is used to point backward to the previous one. Thus a type definition of the following form is required.

```
TYPE termpointer = ^term;
     ^term = RECORD
                   datafieldnames : datatypes;
                   backward,forward : termpointer;
             END;
```

When handling operations such as insertion and deletion of items we must now remember to reconnect the list in both directions.

Thus, for example, if we want to insert a record after the current record we need to break two links and establish four new ones, always remembering to place the new links in position before changing the old ones. This operation is carried out as follows

(* 1. Create a new record *)
    NEW(temporary);
(* 2. Ensure that the forward pointer of this new record points to the next record and then make the forward pointer of the current record point to the new record *)
    Temporaryˆ.forward := currentˆ.forward;
    currentˆ.forward := temporary;
(* 3. Ensure that the backward pointer of the new record points to the beginning of the current record. That is, the record which is going to be the one before it *)
    temporaryˆ.backward := current;
(* 4. Finally, ensure that the backward pointer of the record after the current one points to the beginning of the new record *)
    temporaryˆ.forwardˆ.backward := temporary;

These moves are shown in figure 18.2.

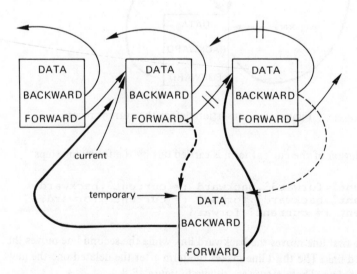

Figure 18.2 Insertion after the current record in a doubly linked list

If the insertion is to be placed before the current term then the steps are

```
NEW(temporary);
(*   establish and move the backward pointers *)
     temporary^.backward := current^.backward;
     current^.backward := temporary;
(*  establish and move the forward pointers *)
     temporary^.forward := current;
     temporary^.backward^.forward :=temporary;
```

This is demonstrated in figure 18.3.

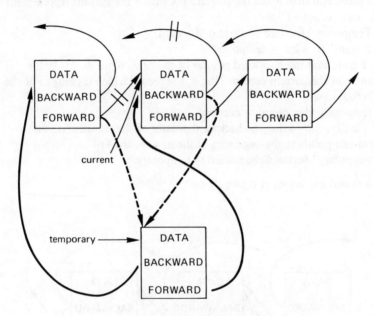

Figure 18.3   Insertion before the current record

Deletion of the current term is carried out by the following steps:

```
current^.forward^.backward := current^.backward;
current^.backward^.forward := current^.forward;
current := current^.forward
```

The first line moves the backward link while the second line moves the forward link. The third line makes the term after the deleted one the new current term. These moves are shown in figure 18.4.

A neat way of working with lists of events is to visualise them as a ring in which the forward pointer at the end of the list points to the start of the list, while the backward pointer at the start of the list points to the end of the list.

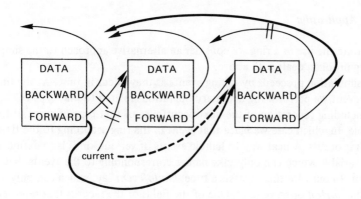

Figure 18.4   Deletion of the current term

This is illustrated in figure 18.5.

Working in this way avoids the need to keep track of NIL pointers at the ends.

As new events are inserted in the ring it expands, while the removal of events causes it to contract.

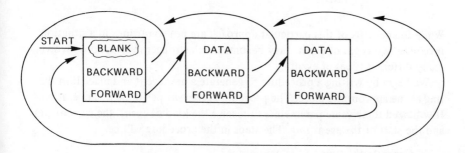

Figure 18.5

When the ring is empty — that is, it contains no events — in order for it to exist we have a dummy entry (record) in which no data is stored. Thus a ring can be initialised by creating an address for a reference pointer which marks the dummy record in the ring and the start of the list. The values of both the forward and backward pointers are then assigned this address. Thus we can initialise a ring as follows (all the variables are of type *termpointer*):

```
NEW(ring);
ring^.forward  := ring;
ring^.backward := ring;
```

18.3.1 Application

To illustrate the use of a ring we consider an alternative approach to the simple queueing problem dealt with above.

The simulation proceeds by taking events as they occur in time. At the time of each event we perform the necessary operations related to that type of event, including generation of the next occurrence of that type of event, if that is possible. In which case we place that event in the ring according to the time at which it occurs. A neat way to handle events of various kinds is to define a type of variable which can only take names corresponding to the events that can occur. In our case this suggests a type, *kindofevent* say, which can only have values *arrival* or *serviceend*. One of the fields of the records that represent each event should be of this type recording the type of event to which that particular record relates.

Thus, for this particular example we need to declare the following types:

```
TYPE kindofevent = (arrival,serviceend);
     eventpointer = ^event;
     event = RECORD
               eventkind : kindofevent;
               time : REAL;
               forward,backward : eventpointer;
             END;
```

While the benefits of this particular approach are not so obvious in a situation involving two types of event, they become much stronger in situations involving many different types of event.

We begin by writing a procedure to generate a new event and place it in the ring at the appropriate place. The procedure *genevent* performs this operation. It will need three value parameters to receive the kind of event, the current time and the start of the event ring. The steps in the procedure will be

> Generate the time to the next event
> Create a new event of the same type
> Store the related information in this new event
> Place an auxiliary event at the start of the ring
> Move backwards round the ring
> > until the time of the new event is greater than the event reached
> Insert the new event in the ring

The last step is performed using the method discussed in the previous section. The only other thing that perhaps needs a comment at this stage is that the time to the next event will probably vary according to the type of event and the simulation problem, which would involve some development of the first step of the above algorithm. The resulting procedure then performs the operation using two local parameters of type *eventpointer* for the creation of the new event and searching for its position in the ring of events. It uses the function *random* described in chapter 15 for which the global INTEGER variable *seed* is used for the initial seed and subsequent values of the sequence of random integers the function produces.

```
PROCEDURE genevent (neweventkind : kindofevent;
                    currenttime,totaltime : REAL;
                    ring : eventpointer);
VAR newevent,currentevent : eventpointer;
    timetoevent:REAL;
BEGIN
(* generate the time till the next event *)
    timetoevent := random;
(* create a new event and assign the data to it *)
    NEW(newevent);
    WITH newevent^ DO
      BEGIN
        eventkind := neweventkind;
        time := currenttime + timetoevent;
      END;
(* find the position of the new event  by moving an
    auxiliary pointer anticlockwise round the ring *)
    currentevent := ring;
    REPEAT
      currentevent := currentevent^.backward;
    UNTIL newevent^.time >= currentevent^.time;
(* place new event in the event ring *)
    newevent^.forward := currentevent^.forward;
    newevent^.backward := currentevent;
    currentevent^.forward^.backward := newevent;
    currentevent^.forward := newevent;
END;
```

We now turn to the main part of the simulation. The basic structure is:

> Initialise the ring
> Generate the time of the first event and indicate that the server is free
> Perform the simulation until the total time is exceeded
> > and the queue is empty (that is, the server is free)

The simulation is carried out by processing the events according to their type. This can be neatly done using a CASE construction such as the following

> CASE eventtype DO
>
> > arrival:
> > > IF server free THEN
> > > > generate the next end of service event
> > > ELSE
> > > > increase queuelength
> > > generate the next arrival

service end:
```
IF queuelength not zero THEN
    generate the next end of service
    decrease the queuelength
ELSE
    indicate that the server is free
```

We complete the simulation of each event by removing the current event from the ring and then moving on to the next event. Thus we need two more steps outside our CASE construction

```
delete the current event from the start of the ring
move on to the next event
```

The deletion of the current event is done by the procedure *delete*. The pointer *currentevent* is then moved to the next event by assigning it the value *ring^.forward* (that is, the new first event in the ring).

The main part of the program follows. The variables required are *ring* and *currentevent* of type *eventpointer* which are used to indicate the start of the ring and the current event respectively. INTEGER variable *queuelength* is used to store the current queuelength while REAL variable *totaltime* contains the length of time the simulation covers. Finally, the BOOLEAN variable *serverfree* is used to indicate the current state of the server.

```pascal
PROGRAM simplequeue(INPUT,OUTPUT);
TYPE kindofevent = (arrival,serviceend);
     eventpointer = ^event;
     event = RECORD
                eventkind : kindofevent;
                time : REAL;
                forward,backward : eventpointer;
             END;
VAR ring,currentevent : eventpointer;
    queuelength,randomseed : INTEGER;
    totaltime : REAL;
    serverfree : BOOLEAN;

FUNCTION random : REAL;
(* function  to generate Pseudo-Random  numbers    see
Chapter 15 *)

PROCEDURE delete(currentevent : ring);
(* procedure to delete the current event *)
BEGIN
  currentevent^.forward^.backward :=
                                 currentevent^.backward;
  currentevent^.backward^.forward :=
                                 currentevent^.forward;
END;

PROCEDURE genevent (neweventkind:kindofevent;
                    currenttime,totaltime:REAL;
                    ring : eventpointer);
```

```
(*  procedure to generate details  of  the  next  event
described above *)

BEGIN
    randomseed := 374; totaltime := 5; queuelength := 0;
(*  create an empty ring *)
    NEW(ring);
    WITH ring^ DO
      BEGIN
        forward := ring;
        backward := ring;
        time := 0;
      END;
(* generate the first arrival *)
    serverfree := TRUE;
    genevent(arrival,0,q);
    currentevent := ring^.forward;

(* perform the simulation *)
    REPEAT
      WITH currentevent^ DO
        CASE eventkind OF

            arrival:
          (* process an arrival *)
              BEGIN
                queuelength := queuelength + 1;
                genevent(arrival,time,totaltime,q);
                IF serverfree THEN
                  BEGIN
                    queuelength := queuelength - 1;
                    genevent(serviceend,time,totaltime,q);
                    serverfree := FALSE;
                  END;
                WRITELN(time:6:2,queuelength:3,
                                            'arrival':10);
              END;

            serviceend:
          (* process a service end *)
              BEGIN
                WRITELN(time:6:2,queuelength:3,
                                            'serviceend':13);
                IF queuelength > 0 THEN
                    BEGIN
                      queuelength := queuelength - 1;
                      genevent(serviceend,time,totaltime,q);
                    END
                  ELSE
                    serverfree := TRUE;
              END;
          END;
      delete(currentevent);
      currentevent := ring^.forward;
    UNTIL (currentevent^.time >= totaltime) AND
                                        serverfree;
END.
```

18.3.2 Handling waiting times

While the above approach handles the sequence of events as each new arrival occurs, information relating to that arrival is lost other than the fact that the queuelength is increased. A more versatile approach would be to leave arrivals in the ring until their service starts.

This can be simply achieved by a minor modification to the procedure *genevent*. We need to include the following lines before we create a new event. The REAL variable *wait* is used to store the time the next person to be served has been waiting in the queue. It should be declared as an additional variable parameter of the procedure.

```
IF neweventkind = serviceend THEN
   BEGIN
      wait := currenttime - q^.forward^.time;
      delete(ring^.forward);
   END;
```

The only other alterations are in the main program when we move on to the next event in time. Now we only perform the deletion of the event from the ring if it is a service end, which affects how we move on to the next event. If the queue is not empty this is the event after the current event, while if the queue is empty this will be the first non-empty event in the ring. Thus the last two lines in the simulation process are replaced by the following three lines:

```
IF currentevent^.eventkind = serviceend THEN
   delete(currentevent);
IF queuelength = 0 THEN
   currentevent := ring^.forward;
ELSE
   currentevent := currentevent^.forward;
```

18.3.3 Concluding remarks

The result of running the programs described in this chapter is to produce a list of events and related information in the case when both the inter-arrival time and service time distributions are uniformly distributed over the interval $(0,1)$. The output shown below covers a period of five time units.

Time	Queue	Event	Waiting time
0.67	0	arrival	
1.12	1	arrival	
1.16	2	arrival	
1.60	2	serviceend	wait = 0.00
1.97	2	arrival	
2.06	3	arrival	
2.07	3	serviceend	wait = 0.48
2.29	2	serviceend	wait = 0.91
2.43	2	arrival	
2.55	2	serviceend	wait = 0.32
2.99	2	arrival	
3.03	2	serviceend	wait = 0.49
3.47	1	serviceend	wait = 0.60
3.81	1	arrival	
3.92	2	arrival	
4.37	2	serviceend	wait = 0.48
4.70	1	serviceend	wait = 0.56
4.70	1	arrival	
4.94	2	arrival	

Once we have a basic program and an understanding of how to handle queues, we can consider various problems such as

(a) the effect of varying the arrival and service time distributions,
(b) construction of the waiting time distribution,
(c) obtaining observations on a server's idle times and constructing an estimate of its distribution,
(d) generalising to the case of several servers.

Using this approach of handling events as they happen and generating information relating to the next event of the same type, we can fairly easily approach more complex problems such as simulating the use of a hospital ward or a bus route.

18.4 Exercises

18.1. Modify the programs to determine the maximum queue-length which occurs over a simulation period. How would you simulate its distribution?

18.2. Modify the programs to estimate the distributions of the server's idle time and the customer's waiting time using the techniques discussed in chapter 14.

18.3. Modify the programs in this chapter to allow for different arrival and service time distributions. For example, allow both of them to have exponential distributions with different parameters. Run simulations to examine what happens if the mean time between arrivals is greater than, equal to or less than the mean length of service. (*Note:* the mean of an exponential distribution is the inverse of its parameter.)

18.4. Modify the programs to allow more than one server. Explore how, for a fixed service time distribution, the number of servers should be increased as the mean time between arrivals decreases.

18.5. A stack can be considered to be a queue in which arrivals are placed at the head, as before, but departures also occur from the head. Write procedures to add and remove items from the stack and to check whether it is empty.

18.6. A post office has three counter positions at which customers are served. The time of service is an exponential variable with mean one minute. Write a program to compare by simulation two alternative queueing systems. In the first, customers join a single queue and as soon as one service position becomes free the person at the front of the queue is served. In the second, three separate queues form, one for each server, each customer joining the shortest at his time of arrival and cannot change queues once he is in the system. Points to examine are the distribution of waiting times of the customers, the maximum number of people waiting at any one time and the idle time of the servers. Experiment with a range of possible times between successive arrivals.

18.7. Write a program to simulate a morning's surgery of a doctor, under two possible options:

(1) no appointments — people turn up as they wish,
(2) fixed appointment system.

Under (1) Assume the first arrival is at 9 a.m. and times between successive arrivals are independent exponential variables with mean 10 minutes, and that consulting times are independent values from a normal distribution with mean 10 minutes and standard deviation 2 minutes.

Under (2) There are fixed appointments every 10 minutes, but patients do not always arrive on time so that successive differences from the appointed time are independent normal values with mean 0 and standard deviation 2.5 minutes. Consulting times are as in (1)

Assume the surgery opens at 9 a.m. and closes at 12 noon.
    Write a program to perform the simulation and supply the following information:

(a) The average time that a patient has to wait.

(b) The maximum number of patients in the waiting room at any given time.

(c) How long the doctor is idle.

In general, does the appointment system appear to offer any advantages (to doctor and patient)? Do you have any criticisms of the model, and what refinements or improvements would you make?

19 Sets

19.1 Introduction

The final structured type we introduce is the SET type which does not occur in
most other programming languages. The SET type defines a structure that can
represent more than one value of a given scalar type. It is defined as follows

```
TYPE settypename = SET of basetype;
```

where *basetype* is the type of the possible values that can be entered in the sets
of the defined type. The *basetype* must be scalar and may be either a user-defined
scalar type, an existing scalar type, such as CHAR, or a subrange of an existing
scalar type. There is, however, a further limitation on the base types we can use.
Most implementations of Pascal only allow a certain number of elements in the
set of possible elements defined by the base type. The maximum number of
elements that a set can contain varies from one implementation of Pascal to
another. However, most modern versions of Pascal allow at least the full character
set of 256 values. Other types allowed must have ordinal values within the range
allowed.

As a result we cannot, for example, define a set type allowing the full range of
possible integers; we are restricted to non-negative integers in the range from 0
to one less than the maximum number of terms allowed in the SET type.

The following examples are valid SET definitions.

```
TYPE   colour = (red,green,blue,black,purple,
                                    yellow,orange);
       colourset = SET OF colour;
VAR hues : colourset;
```

defines a variable *hues* which is a set whose contents can only come from some
or all of the specified colours in the type colour.

```
TYPE letters = SET OF CHAR;
VAR smallletters,capitalletters : letters;
```

defines two variables *smallletters* and *capitalletters* which both represent sets,
the contents of which can only come from the set of characters which make up
the type CHAR.

```
TYPE digits = 0..9
     numberset = SET OF digits
VAR oddnumbers,evennumbers : numberset;
```

defines two variables *oddnumbers* and *evennumbers* which represent sets whose possible elements are members of the set of single digits defined by the subrange type.

However

```
TYPE numbers = SET OF INTEGER;
     wrongnumbers = SET OF (-12..3000);
     largenumbers = SET OF (4356..4375);
```

are not allowed since the maximum size of any set of the first two types is certain to be in excess of the maximum allowed by your implementation of the Pascal language. The second contains negative values which are not allowed. The third is invalid since its range of values almost certainly lies outside the ordinal values allowed.

The set type defines the power set of a scalar type. The power set consists of all possible subsets. If there are b values in the set type, then the power set consists of 2^b possible subsets.

19.2 Assignment to Sets

As usual, when a variable identifier is defined it does not contain any values and cannot be used in subsequent operations until values have been assigned to it. Values may be assigned to sets either as a list of values between square brackets or by assigning another set. Thus

oddnumber := [1,3,5,7,9] ; assigns the values in the list to the set

smallletter := ['a'. . 'z'] ; assigns the characters in the character set from the character 'a' through to and including the character 'z'

evennumber := [ ] ; creates an empty set by assigning a set containing no values to the set *evennumber*

oddnumber := evennumber; sets the contents of the set *oddnumber* equal to the contents of the set *evennumber*

We can also use a variable of the basetype to assign values to a set. The statements

```
number := 2;
evennumber := [number];
```

assign the value currently taken by *number* to the set *evennumber*. The variable *number* must be of the same type as the basetype of the set to which the value is being assigned.

The elements can also be in the form of expressions, provided the result is of the correct type and within the range permitted. Thus, if *i* and *j* are of type *digits* and *numbers* is a set of *digits* — that is, of type *numberset* — then

$$\text{numbers} := [i, (i + j) \text{ MOD } 10, (i - j) \text{ MOD } 10, (i * j) \text{ MOD } 10]$$

is valid. Obviously the use of expressions is restricted to types for which such expressions have a valid meaning.

Finally, it should be noted that the contents of a set are not ordered hence the expressions

```
oddnumber := [1,3,5,7,9];
```

and

```
oddnumber := [3,5,1,9,7];
```

are equivalent.

19.3 Set Operations

As well as assigning sets of values or the contents of other sets, we can also assign the results of the various operations that are allowed on sets. The basic set operations allowed are union, intersection and difference.

The union of two sets is the set of elements that occur in one or both of the two sets. Mathematically, the union of two sets *A* and *B* is represented by $A \cup B$. In Pascal it is denoted by addition of the two sets. Thus

```
oddnumber :=   [1,3,5] + [1,5,7,9]
```

results in *oddnumber* containing values 1, 3, 5, 7, 9.

```
evennumber := evennumber + [number]
```

adds the value of number to the contents of *evennumber*. If the value of *number* is already present, then no change is made to the set *evennumber*.

The intersection between two sets is the set of elements that are contained in both the sets. Mathematically, the intersection between two sets *A* and *B* is represented by $A \cap B$. In Pascal it is denoted by the multiplication of the two sets. Thus

```
oddnumber := [1,3,5] * [3,5,6,7]
```

results in *oddnumber* containing the values 3 and 5 which are the only values that occur in both sets.

The difference between two sets is the set of elements in the first set that do not occur in the second. Thus

```
oddnumber := [1,2,3,4,5,6,7,8,9,0] - [2,4,6,8,0]
```

results in *oddnumber* containing the values 1, 3, 5, 7, 9.

As well as these operations which can be performed on sets, there are also several BOOLEAN operators that apply to sets. These are

=   equality of sets
> thus $a = b$ is TRUE if the sets a and b are identical

⟨⟩   inequality of sets
> thus $a ⟨⟩ b$ is TRUE if no member of the set a is in the set b

<=   contained in
> thus $a <= b$ is TRUE if the whole of set a is within the set b
> ($A \subset B$ in mathematical notation)

>=   contains
> thus $a >= b$ is TRUE if the whole of set b is in the set a

IN   member of
> thus v IN a is TRUE if the element v is contained in the set a
> ($v \in A$ in mathematical notation)

The relations *contained in* and *member of* are very closely related. In the usual mathematical notation we say the element x is a member of the set (x, y) — that is, $x \in (x, y)$ — while the set (x) is contained in the set (x, y) — that is, $(x) \subset (x, y)$. The equivalent statements in Pascal are x IN $[x, y]$ and $[x] <= [x, y]$ respectively.

We can also make use of the IN relation to avoid constructing complex conditions. With any of the scalar types we can examine whether, at any particular time, an identifier of the appropriate type takes a value within a subset of the values permitted for that type.

Thus, for example, in our program *graphexample* described in chapter 13 we used the following construction

> acceptableanswer := (ans = 'S') OR (ans = 's')
>   OR (ans = 'F') or (ans = 'f');
> IF NOT acceptableanswer THEN etc . . . .
>   . . . .
> UNTIL acceptableanswer;

Using the IN relation we can avoid using the BOOLEAN variable *acceptableanswer* and simply check whether the value of *ans*, of type CHAR, was in the se ('S', 's', 'F', 'f'). Thus, for example

```
ans  IN  ['S','s','F','f']
```

could be used in the UNTIL statement. The statement is true if the variable *ans* currently has one of these four values assigned to it, otherwise it is false.

Similarly

```
IF  NOT(ans IN ['S','s','F','f']) THEN ....
```

could be used in the IF statement.

Note here that the NOT does not go immediately before the IN but goes before the element name. Thus, ans NOT IN ['S', 's', 'F', 'f'] is invalid.

19.4   Programming with Sets

The problem of entering values into a set depends on the type of the set's elements. If the base type is subrange INTEGER or CHAR, then the values can be read into a variable which is then added to the set.

For example

```
READ(element);
labels := labels + [element];
```

would add an element of these two base types to an existing set, *labels*, of the appropriate type.

However, if the type of the set's elements is a user-defined type, then the READ must be adapted by use of the CASE statement to convert readable types, INTEGER and CHAR, to the user-defined type as described in section 13.2.2.

If we wish to display the contents of a set, then we must check each value in the base type and display only those present in the set. Thus, if we have a set *labels* of subrange INTEGER or CHAR, then we can display the contents of the set using

```
IF labels = [] THEN WRITELN('the set is empty');
ELSE
   FOR element := firstelementintype TO
                                  lastelementintype DO
     IF element IN labels THEN  WRITE(element);
```

where *firstelementintype* and *lastelementintype* are the first and last elements in the ordered list of the base type's values.

If the base type is a user-defined type, then the method of displaying an element's value must be modified by use of a case statement to convert the element's name into a quantity that can be displayed (see section 13.2.2).

The number of elements in a set is called its *cardinality* and can be obtained by checking whether each element in the base type is a member of the set, adding 1 to a count of the number of elements if it is present. The following function will determine the cardinality of a set *labels* of type *names*. The parameters *firstelementintype* and *lastelementintype* will receive the first and last elements in the base type's range.

```
FUNCTION cardinality(labels : names;
                     firstelementintype,
                     lastelementintype : basetype) : INTEGER;
VAR count : INTEGER; element : basetype;
BEGIN
   count := 0;
   FOR element := firstelemenintype TO
                                    lastelementintype DO
     IF element IN labels THEN count := count + 1;
   cardinality := count;
END;
```

To find the number of a particular element in a set we again search through all the elements in the base type, counting those elements in the set until the required element is reached. The following function returns the number of an element:

```
FUNCTION elementnumber(labels : names; element,
               firstelementintype : basetype) : INTEGER;
VAR i : INTEGER; e : basetype;
BEGIN
   IF element IN labels THEN
     BEGIN
       i := 0;
       e := firstelementintype;
       IF e IN labels THEN i := i+1;
       WHILE e <> element DO
         BEGIN
           e := succ(e);
           IF e IN labels THEN i := i+1;
         END;
       elementnumber := i;
     END
   ELSE
     elementnumber := 0;
END;
```

The function returns the value zero if the element is not in the set.

The final function in this section finds the element in the set that has a value from *firstelementinsearch* onwards in the order of the base type. If no element is found, then *findelement* has the value of *firstelementinsearch* and that element is not a member of the set.

```
FUNCTION findelement(currentset : names;
        firstelementinsearch : basetype) : basetype;
VAR y : basetype;
BEGIN
  y := firstelementinsearch;
  WHILE NOT(y IN currentset) DO
    y := succ(y);
  findelement := y;
END;
```

19.5   Binary Relationships

It is useful to conclude this chapter by considering relationships between members of sets and how we might represent such relationships on a computer. We will use these in the final chapter as a means of representing paths existing in a graph.

A *binary relation* R between two sets A and B defines a subset of the product set $A \times B$. That is, it defines a set of ordered pairs from the complete set of ordered pairs $A \times B$.

If $A = B$, then we say that the relation is on A.

If $x \in A$ and $y \in B$, then if x is related to y we use the notation xRy.

One way of representing a relation R on a set is by a logical matrix. Thus if we have sets

$$A = [a_1, a_2 \ldots a_n]$$
$$\text{and} \quad B = [b_1, b_2 \ldots b_k]$$

then the logical matrix M is given by

$$M[i, j] = \frac{\text{FALSE if } (a_i, b_j) \notin R}{\text{TRUE \ \ if } (a_i, b_j) \in R}$$

This matrix defines a subset of the product set.

Thus, for example, if $A = (a,b,c)$ and $B = (x,y,z)$ and

$$M = \begin{bmatrix} T & T & F \\ T & F & T \\ F & F & T \end{bmatrix}$$

then the relationship R consists of the ordered pairs

$$[(a,x), (a,y), (b,x), (b,z), (c,z)]$$

and is illustrated in figure 19.1.

There are three basic properties of a relation R on a set A which may be of interest. These are

reflexive if for all $x \in A$, xRx
symmetric if for all $x,y \in A$, $xRy \Rightarrow yRx$
transitive if for all $x,y,z \in A$, xRy and $yRz \Rightarrow xRz$

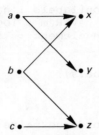

Figure 19.1 An example of a binary relationship

Figure 19.2 illustrates examples of relationships satisfying these three properties and their corresponding logical matrices. Note that for the reflexive example the diagonal terms are all true, while for the symmetric case the matrix is also symmetric.

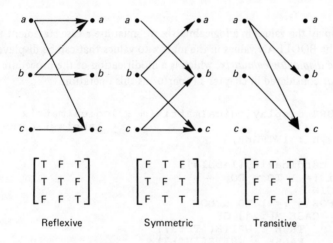

$$\begin{bmatrix} T & F & T \\ T & T & F \\ F & F & T \end{bmatrix} \qquad \begin{bmatrix} F & T & F \\ T & F & T \\ F & T & T \end{bmatrix} \qquad \begin{bmatrix} F & T & T \\ F & F & T \\ F & F & F \end{bmatrix}$$

Reflexive   Symmetric   Transitive

Figure 19.2

One way of representing a logical matrix on a computer is by a two-dimensional array of type BOOLEAN. Thus, if *max* represents the maximum cardinality of any set we consider, we can declare a type *logicalmatrix* as follows

```
TYPE logicalmatrix = ARRAY [1..max,1..max] of BOOLEAN
```

We begin by considering how to create the values in such a matrix. The simplest way is to enter the ordered pairs present in the relationship and construct the matrix from these. One way of doing this is shown in the procedure *enterrelation*.

```
PROCEDURE enterrelation(labels : names;
                              VAR m : logicalmatrix);
VAR firstelement,secondelement : keytype;
    i,j,n : INTEGER;
BEGIN
   n := cardinality(labels);
   FOR i := 1 TO n DO
     FOR j := 1 TO n DO
       m[i,j] := FALSE;
   WRITELN('enter the ordered pairs of the relation ');
   REPEAT
     READ(firstelement,secondelement);
     IF (firstelement IN labels) AND
                        (secondelement IN labels) THEN
       BEGIN
         i := elementnumber(labels,firstelement);
         j := elementnumber(labels,secondelement);
         m[i,j] := TRUE;
       END;
   UNTIL NOT(firstelement IN labels);
END;
```

To display the values in a logicalmatrix we must use a case statement to convert the BOOLEAN values in the matrix to values that can be displayed. Procedure *displaylogicalmatrix*, which is a modification of the procedure *displaymat* considered in chapter 11, performs this operation.

```
PROCEDURE displaylogicalmatrix(m : logicalmatrix;
                                 labels : names);
VAR i,j,n : INTEGER;
BEGIN
   n := cardinality(labels);
   FOR i := 1 TO n DO
     BEGIN
       FOR j := 1 TO n DO
         CASE M[i,j] OF
            TRUE : WRITE('T':3);
            FALSE : WRITE('F':3);
         END;
       WRITELN;
     END;
END;
```

19.5.1 Convolution of relations

If we have three sets

$$A = [a_1, a_2 \ldots a_n],$$
$$B = [b_1, b_2 \ldots b_q] \text{ and}$$
$$C = [c_1, c_2 \ldots c_p]$$

and relations R from A to B and S from B to C, then the composite relation from A to C is given by $S \circ R$. If R is described by the logical matrix M_1 and S is described by the logical matrix M_2, then the composite relation $S \circ R$ is described by the logical matrix product

$$M_3 = M_1 \cdot M_2$$

in which $a_i S \circ R \, c_j$ if $\exists k$ such that $a_i R b_k$ and $b_k S c_j$.

Hence

$$
\begin{aligned}
M3[i, j] = &(M1[i, 1] \text{ AND } M2[1,j]) \text{ OR} \\
&(M1[i, 2] \text{ AND } M2[2,j]) \text{ OR} \\
&\ldots\ldots\ldots\ldots\ldots\ldots \text{ OR} \\
&(M1[i, q] \text{ AND } M2[q, j])
\end{aligned}
$$

Thus $M3[i, j]$ is true if there is at least one pair $M1[i, k]$ and $M2[k, j]$ for which both are true. Thus if M1, M2 and M3 are declared to be **ARRAY** [1..max, 1..max] of BOOLEAN we can program this as

```
FOR i := 1 TO n DO
  FOR j := 1 TO p DO
    BEGIN
      k := 0;
      REPEAT
        k := k + 1;
        M3[i,j] := M1[i,k] AND M2[k,j];
      UNTIL (k = q) OR M3[i,j];
```

Note that we only proceed with the calculation up to the first instance of both $M1[i, k]$ and $M2[k, j]$ being true.

The application of some of these ideas will be explored in the final chapter.

19.6 Exercise

19.1. Write procedures to check whether a relationship is symmetric, reflexive and/or transitive.

20 Trees and Graphs

20.1 Introduction

A travelling salesman lives in Doncaster and has to visit Aberdeen, Bradford and Cardiff. If he knows the distances between each of the four towns, what is the shortest route which ensures that he visits each town once only and ends up at his starting point? Figure 20.1. shows the distances between the towns.

There are only three distinct tours that he can make namely DABCD (distance 1061), DCABD (1040) and DBCAD (1071). Thus the shortest route is Doncaster-Cardiff-Aberdeen-Bradford-Doncaster.

As the number of towns is increased, then the number of alternative routes increases rapidly and solving the problem using this approach, of looking at all possible routes, soon becomes unmanageable even using a computer. Is there a quicker alternative method for producing a solution or at least a close approximation to the solution?

A telephone engineer has the same information as shown in figure 20.1. His problem, though, is to plan a telephone network that will connect all the four towns using the minimum amount of cable.

His solution is to start with the shortest distance and then add in the smallest distances of those remaining, provided the inclusion of a connection does not form a cycle with those already selected, until all towns have been connected.

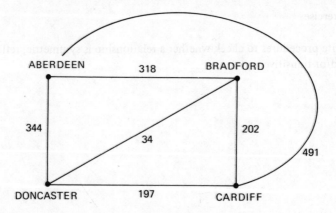

Figure 20.1   Distances between four towns

Thus he starts with Doncaster to Bradford (distance 34) and Doncaster to Cardiff (197). The next smallest is Bradford to Cardiff (202) but this would create a loop or cycle with the existing connections. He therefore ignores this connection and includes Bradford to Aberdeen to complete the process, since every town is now included in the set of connections. Figure 20.2 illustrates the connections that have been chosen.

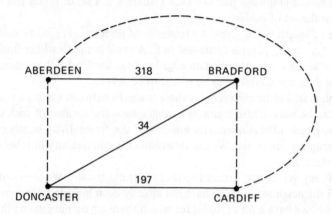

Figure 20.2   Minimum connection between four towns

This is known as the minimum connection problem, which in general is posed in terms of finding the shortest distance that can be used to link all the towns if we know the distances between some or all of the towns. How can we write a computer program to solve this in the general case of *n* towns?

The description and solution of these and related problems fall into the study of trees and graphs. An easy-to-read introduction to these areas is given by Skvarcius and Robinson (1986) while a very good, though more advanced introduction, is given by Wilson (1985). It is an area of mathematics in which we can exploit a wide range of features of the Pascal language and in particular the SET and POINTER structures.

20.2   Undirected Graphs

A *graph* consists of a set of vertices or nodes together with a set of edges or lines which connect the nodes. Figure 20.1 is an example of a graph.

A graph is either *directed*, in which case the lines or edges also have directional information (for example, arrows) indicating the direction in which we can pass from one node to the next, or it is *undirected*, in which case the edge is simply treated as a join between nodes and we can pass along it in either direction.

Expressed in these terms, graph theory can be thought of as an extension of the idea of a relation on a set considered in chapter 19, since we can think of the edge between two nodes as an ordered pair. Hence any simple graph can be described in terms of a set of nodes and a related adjacency matrix which defines a relation on the node set. If the graph is undirected the adjacency matrix is symmetric.

Thus a simple graph is a pair $G = (V,E)$ where V is a set of nodes and E is a relation on the set of nodes.

A *path* of length k in a graph is a sequence of nodes $v_0, v_1, \ldots v_k$ such that for all $i = 1 \ldots k$, (v_{i-1}, v_i) is contained in E. A *cycle* is a path whose final vertex is the same as the first vertex and no edge has been used more than once. A graph that does not contain a cycle is said to be *acyclic*.

A graph is said to be *connected* if there is a path between every pair of nodes. A graph can be broken into connected subgraphs — the number of such subgraphs being called the *connectivity number*. If the connectivity number is 1, then the graph is connected. We can determine the connectivity number using the next algorithm.

Start at any vertex and proceed to find all its neighbours. We then proceed to find all the neighbours which have not already been included. We repeat this process until we have a set of nodes for which there are no neighbours that have not already been included. The resulting set is one component of the graph. Selecting a vertex that has not already been included, we jump to the next component and repeat the whole process until no more components can be found. The number of separate components found is the connectivity number. Thus, if we start with a set that contains all the nodes in the graph, and its adjacency matrix, m, we can proceed as follows

> Assign the value zero to a counter
> WHILE the set of nodes is not empty
>> find the first element in the set
>> find the set of elements connected to that element
>> add one to the counter
>> remove the set of connected elements from the set of nodes
> connectivity number is the final value of the counter

The main problem is to find the set of elements connected to the first element. The first element in a set of nodes can be found using the function *findelement* given in chapter 19.

A set of connected elements can be determined as follows, starting from a set *currentset* which at the start contains all the nodes. Starting with the first element in this set, we then proceed to find all the elements connected to that element. The set, *connectedelements*, will contain the set of elements connected to the first. The adjacency matrix enables us to find the nodes directly linked to our first node, however we also have to find the nodes linked to these and so on. We therefore create a third set, *newinset*, in which we store the elements to which we have not previously established connections. If this is not empty, we must establish all the connections to these elements which can be done recursively.

The whole process can be achieved using the following procedure and function, the former using the function *findelement* described in chapter 19. Note that 'a' and 'z' represent the first and last members in the set type *names* used in this example.

```
PROCEDURE connections(m : logicalmatrix;
                      labels,currentset : names;
                      VAR connectedelements : names);
VAR y,j : keytype; newinset : names;

BEGIN
(* at the outset currentset contains all the nodes to
   which we we want to establish connections *)
  WHILE currentset <> [] DO
    BEGIN
(*   create   an   empty   set   for   receiving   any   newly
     connected  nodes  and  find  the  first  element  in  the
     currentset of nodes *)
       newinset := [];
       y:=findelement(currentset,'a');

(* remove  this  element   from  the  current  set  and  if  it
   is not yet in the set of connected elements place it
   in the set of connected elements *)
       currentset := currentset - [y];
       IF connectedelements = [] THEN
         connectedelements := connectedelements + [y];
       IF y IN connectedelements THEN
         BEGIN
             IF y <> 'z' THEN
             BEGIN

(* find the set of  elements  for  which  we  have  not
   established whether thay are connected  which can be
   reached from this element *)
             FOR j := succ(y) TO 'z' DO
                IF j IN labels THEN
                   IF NOT(j IN connectedelements) AND
                      m[elementnumber(labels,y),
                              elementnumber(labels,j)] THEN
                      newinset := newinset + [j];

(* add these new connections to the set of connections
   and  find the subsequent connections from these
   elements *)
```

```
            IF newinset <> [] THEN
               BEGIN
                  connectedelements := connectedelements
                                       + newinset;
                  connections(m,labels,newinset,
                                       connectedelements);
               END;
            END;
   (* remove the new connections from  the  set  of nodes
   whose connections we have still to establish *)
            currentset := currentset - newinset;
         END;
      END;
END;
```

This procedure can then be called from the procedure *connectivitynumber* which determines the number of linked sets that exist.

```
FUNCTION connectivitynumber(m : logicalmatrix;
                           labels : names) : INTEGER;
VAR connectedelements : names;
    count : INTEGER;

BEGIN

   count := 0;
   WHILE labels <> [] DO
      BEGIN
         connectedelements := [];
         connections(m,labels,labels,connectedelements);
         count := count + 1;
         labels := labels - connectedelements;
      END;
   connectivitynumber := count;

END;
```

20.2.1 Hamiltonian circuits and travelling salesmen

A *Hamiltonian circuit* is a circuit that passes through each node exactly once. Such a circuit is a solution to the travelling salesman problem we posed in the introduction. This is a problem of determining a circuit of towns (the nodes) when every town is linked to every other town by an edge. Each edge has a positive weight associated with it, representing the distance between the towns. The problem is to determine a tour of all the towns, ending up at the starting place, which involves the minimum amount of travelling. Such a tour is a Hamiltonian circuit. As we saw in the introduction, one solution is to determine the weight of all the possible Hamiltonian circuits and select the one with the minimum. It should be reasonably clear, however, that this will result in considerable effort as the number of nodes increases. Figure 20.1 shows a graph of four nodes while the three Hamiltonian circuits that can be made are shown in figure 20.3. The last is the one with minimum weight and hence is the solution in this case.

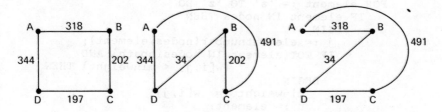

Figure 20.3   Three Hamiltonian circuits in figure 20.1

It has in fact proved impossible to devise an algorithm that produces the optimal solution to this problem within a reasonable amount of time. However, the algorithm given below is one which will produce a suboptimal solution. That is one that is reasonably close to the optimal solution.

> select any node and call it *nodeone*
> assign *nodeone* to *latestnode* and
> > use it to start a set of nodes in the path
> set the total weight to zero
> WHILE there are nodes that are not yet in the path
> > find the node which is not yet in the path
> > > that is closest to the lastest node
> > include the new node in the path and make it the current one
> > add its weight to the total weight
> add the first node to the path to complete the circuit

The procedure *subopt* given below carries out this process.

```
PROCEDURE subopt(nodes :names; w : weights;
                          VAR totalweight : INTEGER);
VAR nodeone,latestnode,u,element : keytype;
    nodesinpath : names;
    minweight,i,j : INTEGER;

BEGIN

  nodeone := findelement(nodes,'a');
  latestnode := nodeone;
  totalweight := 0;
  nodesinpath := [latestnode];
```

```
   WHILE nodesinpath <> nodes DO
     BEGIN

       minweight := 1000;
       i := elementnumber(nodes,latestnode);

       FOR element := 'a' TO 'z' DO
          IF element IN nodes THEN
             BEGIN
                j := elementnumber(nodes,element);
                IF NOT(element IN nodesinpath) AND
                            (w[i,j] < minweight) THEN
             BEGIN
                minweight :=  w[i,j];
                u := element;
             END;
          END;

       WRITELN(u:3,minweight);
       nodesinpath := nodesinpath + [u];
       totalweight := totalweight + minweight;
       latestnode := u;

     END;

   WRITELN(nodeone:3);
END;
```

Since the resulting set of nodes in the path is not ordered, it does not contain details of the final path. Simply to display the path we could output the name of each new vertex as it is selected, as shown in the procedure *subopt*. However, if we need to use the path later, we must also consider ways of storing its order. In Pascal we have a choice of two ways as to how we do this. We can either store the node names in an array, in the order that they are selected, or use a linked list of records as discussed in chapter 17. The latter choice has the advantage that it is more flexible, the list being as long as the cardinality of the set of nodes, whereas the array length must be equal to the maximum cardinality of all the sets that we are likely to use.

20.3 Trees

A *tree* is a connected acyclic graph. An example is shown in figure 20.4.

A tree also has the properties:

1. There is exactly one path between any pair of nodes.
2. The number of edges will be one less than the number of nodes.

Other terminology related to trees:

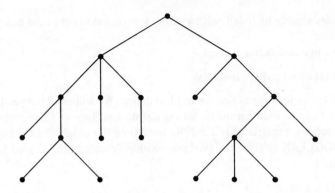

Figure 20.4 A tree

Its *root* is the first node. Each node is a *parent* while the nodes it points to are called its *children*. The connection between nodes is called a *branch* while a node that has no children is called a *leaf*.

We shall see that the tree is a useful vehicle for handling hierarchical relationships and in particular it lends itself naturally to the problems of sorting and searching.

The solution to the minimum connection problem posed in the introduction is an example of a tree. In this case it is called a *spanning tree* since it connects all the nodes in the graph without having a cycle present. It can be shown that every connected graph has a spanning tree. If the edges have weights, we can search for the minimal spanning tree — that is, the one with the least weight. The minimum connection problem's solution, therefore, is the minimum spanning tree for the given graph. A minimal spanning tree can be derived in several ways. Two algorithms which allow us to do this are those by Kruskal and Prim.

Kruskal's Algorithm is

1. Choose any edge which has minimal weight.
2. Choose the remaining $n-2$ edges by choosing the edge with the smallest weight from those remaining such that its inclusion does not form a cycle.

It was this algorithm that we used to determine our solution to the problem in the introduction. However, from a programming point of view we consider Prim's algorithm.

Prim's algorithm is

1. Place a node, v_1 say, into a set P.
   Define set $N = V - (v_1)$ where V is the set of all nodes.
2. After i steps $P = (v_1 \ldots v_i)$ and $N = V - P$
   At this stage we find the shortest edge that connects a vertex x in P, the set

of nodes already included, with a vertex y in N, the set of nodes not yet included.

Place y in P and delete it from N.

Step 2 is repeated until N is empty.

There are various types that we need to define. We will need to specify the names that can be used for the nodes and define a set type of these names. We will also require a matrix of INTEGERs to contain the weights of the edges and one of BOOLEAN to store the final relationship describing the chosen connections.

```
TYPE basetype = 'a' .. 'z';
     names = SET OF basetype;
     logicalmatrix = ARRAY[1..20,1..20] OF BOOLEAN;
     weights = ARRAY[1..20,1..20] OF INTEGER;
```

The following procedure *primtree* is a translation of the algorithm into Pascal which uses the function *findelement* described in chapter 19 to determine the number of each element in the set of nodes. It also uses the function *cardinality* to determine the number of elements in that set.

```
PROCEDURE primtree(v : names; W : weights;
                   VAR totalweight : INTEGER;
                   VAR M : logicalmatrix);
VAR present,absent : names;
    nextnode,nodepresent,nodeabsent : keytype;
    nextstart,nextend,i,j,n : INTEGER;
    minweight : INTEGER;
BEGIN

  totalweight := 0;
  n := cardinality(v);
  FOR i := 1 TO n DO
    FOR j := 1 TO n DO
      m[i,j] := FALSE;
  nextnode := findelement(v,'a');
  present := [nextnode];
  absent := v - [nextnode];

(* while there are still nodes to visit find the
   smallest weight joining an element in set of nodes
   so far included to an element in the set of nodes
   not yet included and and record the start and
   finish of the edge. *)

  WHILE absent <> [] DO
    BEGIN
      minweight := 1000;
      i := 0;
```

```
      FOR nodepresent := 'a' TO 'z' DO
        IF nodepresent IN present THEN
          BEGIN
            i := elementnumber(v,nodepresent);
            j := 0;
            FOR nodeabsent := 'a' TO 'z' DO
              IF nodeabsent IN absent THEN
                BEGIN
                  j := elementnumber(v,nodeabsent);
                  IF minweight > w[i,j] THEN
                    BEGIN
                      minweight := w[i,j];
                      nextstart := i;
                      nextend := j;
                      nextnode := nodeabsent;
                    END;
                END;
          END;
```

```
(*   Insert the edge in the logical  matrix   adding the
     smallest  weight  to  the  totalweight.   Add the
     next node to the set of those included  and remove
     it from the set of those not yet included. *)

      m[nextstart,nextend] := TRUE;
      totalweight := totalweight + minweight;
      present := present + [nextnode];
      absent := absent - [nextnode];
      WRITELN(nextnode);
    END;
END;
```

20.3.1 Binary trees

A special case of a tree that has important uses is a binary tree in which each
node only has at most two branches from it. An example is shown in figure
20.5.

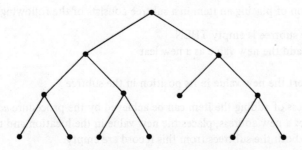

Figure 20.5   A binary tree

A binary tree can be stored using records of pointer type having just two pointer fields, right and left, which point to the two branches from that record. Thus we can set up a type *binarytree* as follows:

```
TYPE binarytree = ^node;
     node = RECORD
               datafields : datatypes;
               left,right  : binary tree;
            END;
```

As usual, there may be one or more datafields each with their corresponding datatypes. For the purpose of illustration we will have a single datafield called *name* or *basetype* in which we will store the name of the node.

The operations that we need to consider for such a structure are those of creating the structure, inserting new items in the structure, searching for items in the structure and displaying the data stored in the structure. Examining figure 20.4, we see how lower levels of the tree are nested in the nodes of higher levels, hence this is a structure that lends itself to the use of recursive programming which we discussed briefly in chapter 7.

We will begin by considering how to insert items since this will also be used in the creation of the data structure.

An item is inserted in the tree structure by creating a new leaf in the position relating to the new item's position in the data's order. We will write the procedure in terms of ascending order of magnitude but other orderings may be substituted.

We start the process of insertion by starting at the root of the tree, then at each node reached the following steps are carried out.

> IF the new value precedes the current node value THEN
> >   place the new value in the left subtree
> ELSE
> >   IF the new value comes after the current node value THEN
> > >   place it in the right subtree
> >   ELSE
> > >   the new value is already in the tree

The operation of placing an item in a subtree consists of the following

> IF the subtree is empty THEN
> >   add the new value as a new leaf
> ELSE
> >   insert the new value in its position in the subtree

The process of adding the item can be achieved by the procedure *additem* which creates a new address, places the new value in the location and then ensures that both the subtrees from this record are empty.

```
PROCEDURE additem(VAR current : binarytree;
                      newvalue : basetype);
BEGIN
  NEW(current);
  current^.name := newvalue;
  current^.left := NIL;
  current^.right := NIL;
END;
```

Inserting the new value in its position in the subtree is the same as the original insertion, hence it can be achieved by calling the procedure recursively. The procedure *insert* will perform the operation of placing a new item in its appropriate position in the data structure.

```
PROCEDURE insert(newvalue : basetype; current : binarytree);
BEGIN
  IF newvalue < current^.name THEN
    IF current^.left = NIL THEN
      additem(current^.left,newvalue)
    ELSE
      insert(newvalue,current^.left)
  ELSE
    IF newvalue > current^.name THEN
      IF current^.right = NIL THEN
        additem(current^.right,newvalue)
      ELSE
        insert(newvalue,current^.right)
    ELSE
      WRITELN('item already in tree ');
END;
```

To create a tree, we simply create the first item using the procedure *additem* and insert the remaining items that are to be included in the structure. This can be achieved, for example, by the procedure *create* which uses data values already stored in a vector *data*.

```
PROCEDURE  create(VAR  tree  :  binarytree;
                  data : vector; noofterms : INTEGER);
VAR  i : INTEGER;
BEGIN
  additem(tree,data[1]);
  FOR i := 2 TO noofterms DO
      insert(data[i],tree);
END;
```

An in-order tree traversal consists of visiting each node in order. Thus, for each node in turn, starting from the root, we visit all the nodes in the left subtree followed by the node itself and then all the nodes in the right subtree. The first and last visits can only be performed if the subtrees are not empty. Thus we have three steps

> IF the left subtree is not empty THEN visit the left subtree
> visit the node
> IF the right subtree is not empty THEN visit the right subtree

Since the steps in the visit to a subtree are the same as the above, then a procedure that traverses a tree can be very simply programmed using the recursion facility. The procedure *printtree* illustrates such a traversal for printing the names of the nodes in ascending order.

```
PROCEDURE printtree(tree : binarytree);
BEGIN
   IF tree^.left <> NIL THEN
            printtree(tree^.left);
   WRITE(tree^.name);
   IF tree^.right <> NIL THEN
            printtree(tree^.right);
END;
```

The in-order traversal also allows us to take a vector of data values and create a second vector, *sorteddata* say, of the values sorted in ascending order. This is done using the following two procedures together with those already described.

```
PROCEDURE sort(tree : binarytree; VAR n : INTEGER);
BEGIN
   IF tree^.left <> NIL THEN sort(tree^.left,n);
   n := n+1;
   sorteddata[n] := tree^.name;
   IF tree^.right <> NIL THEN sort(tree^.right,n);
END;

PROCEDURE sortdata (VAR sorteddata : vector; data : vector;
                        noofobservations : INTEGER);
VAR   n : INTEGER;
      tree : binarytree;
BEGIN
   create(tree,data,noofobservations);
   n := 0;
   sort(tree,n);
END;
```

This is a simple sorting algorithm. Clearly, it is not optimal. At worst there may be $n-1$ levels relating to n data points if the data to be entered into the tree structure is already in order or reverse order. An ideally balanced tree would have k levels, where $2^k < n < 2^{k+1}$. That is, $k = \log_2 n + 1$.

The final operation that we will consider on binary trees is that of searching to see whether a particular item is present in the tree. Suppose we have a value in the variable *searchvalue* which we want to check is in the tree, then we can proceed as follows, starting again at the root.

> IF current node name is the searchvalue THEN search successful
> ELSE
> > if searchvalue is less than the currentnode name AND
> > > the left subtree is not empty THEN
> > > > search the left subtree
> > ELSE
> > > if the search value is greater than the current node name AND
> > > > THE RIGHT SUBTREE IS NOT EMPTY THEN
> > > > > search the right subtree
> > > ELSE
> > > > the search value is not in the tree

The result of this is the recursive procedure *searchtree*:

```
PROCEDURE searchtree(searchvalue : basetype;
                     current : binarytree);
BEGIN
  IF current^.name = searchvalue THEN
           WRITELN('value found')
  ELSE
      IF (searchvalue < current^.name) AND
                  (current^.left <> NIL) THEN
          find(searchvalue,current^.left)
        ELSE
        IF (searchvalue > current^.name) AND
                  (current^.right <> NIL) THEN
            find(searchvalue,current^.right)
          ELSE WRITELN('value not found');
END;
```

20.4 Directed Graphs

In a directed graph $G = (V,E)$, the order of the pair in the relation E gives the direction in which we can move from one node to another. Thus, if $(u,v) \in E$, then we can move from u to v. However, we cannot move from v to u unless (v,u) is also present in E. Thus, the adjacency matrix which describes the relationship E is not symmetric and defines all the paths of length one in the graph.

Paths of length two are represented by the compound relation $E^2 = E \circ E$ which consists of all ordered pairs which are reachable by some intermediary node. Thus, in general $E^k = E \circ E \circ \ldots \circ E \circ E$; the kth compound of E will describe the ordered pairs that are reachable via $k-1$ intermediary nodes. In the previous chapter we saw that E^2 can be represented by M^2 and hence E^k

is represented by M^k. Thus, if there are n nodes in the set and M is the adjacency matrix describing E, then v is reachable from u if (u,v) is in

$$E \text{ OR } E^2 \text{ OR } E^3 \text{ OR } \ldots \text{ OR } E^n$$

Thus $M^* = M \text{ OR } M^2 \text{ OR } \ldots \text{ OR } M^n$ is the matrix that shows which nodes are reachable from which.

Using the algorithm and types described at the end of the previous chapter, we can write a procedure *convolve* which calculates the adjacency matrix for the convolution of two relationships.

```
PROCEDURE convolve(m1,m2 : logicalmatrix;
                   VAR m3 : logicalmatrix;
                   labels : names);
VAR n,i,j,k : INTEGER;
BEGIN
  n := cardinality(labels);
  FOR i := 1 TO n DO
    FOR j := 1 TO n DO
        BEGIN
          k := 1;
          REPEAT
            m3[i,j] := m1[i,k] AND m2[k,j];
            k := k+1;
          UNTIL (k > n) OR m3[i,j];
        END;
END;
```

Once we have this procedure we can calculate the adjacency matrix M^*, which gives us all the paths that exist, as follows:

```
n := cardinality(labels);
m2 := m1;
mstar := m1;
FOR p := 2 TO n DO
  BEGIN
    convolve(m1,m2,m3,labels);
    m2 := m3;
    FOR i := 1 TO n DO
      FOR j := 1 TO n DO
        mstar[i,j] := mstar[i,j] OR m3[i,j];
  END;
```

If we have a relation described by the adjacency matrix, M, where

$$M = \begin{bmatrix} F & T & F & F \\ F & F & T & F \\ T & F & F & F \\ T & F & T & F \end{bmatrix}$$

then

$$M^* = \begin{bmatrix} T & T & T & F \\ T & T & T & F \\ T & T & T & F \\ T & T & T & F \end{bmatrix}$$

This is illustrated in figure 20.6. The paths reachable in two, three and four steps are also shown in figure 20.6, while the matrix M^* shows all the paths that exist.

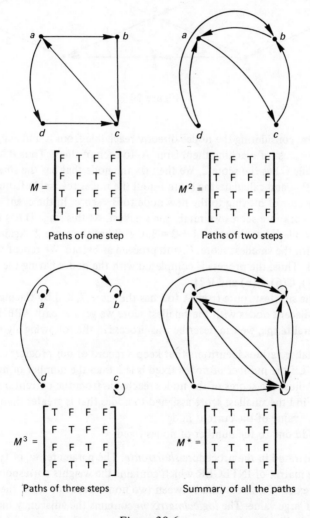

$$M = \begin{bmatrix} F & T & F & F \\ F & F & T & F \\ T & F & F & F \\ T & F & T & F \end{bmatrix}$$

Paths of one step

$$M^2 = \begin{bmatrix} F & F & T & F \\ T & F & F & F \\ F & T & F & F \\ T & T & F & F \end{bmatrix}$$

Paths of two steps

$$M^3 = \begin{bmatrix} T & F & F & F \\ F & T & F & F \\ F & F & T & F \\ F & T & T & F \end{bmatrix}$$

Paths of three steps

$$M^* = \begin{bmatrix} T & T & T & F \\ T & T & T & F \\ T & T & T & F \\ T & T & T & F \end{bmatrix}$$

Summary of all the paths

Figure 20.6

20.4.1 Shortest routes

Another problem that we can consider, which can be represented in terms of a
directed graph, is that of finding the shortest route across a network.

Consider, for example, the network shown in figure 20.7 in which the
numerical values represent the cost of using that particular path.

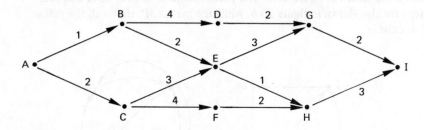

Figure 20.7

We begin by considering the nodes directly reachable from A and assign a score,
equal to the cost of reaching them from A, to each of them. Thus B has the
score 1 while C has the score 2. We then fix the node that has the smallest score
— that is, B — and calculate the score for all the nodes reachable from B by
adding the cost of moving to the new node to B's score. In the event of a node
having two or more scores we retain the smallest. At this stage D has the score
$1 + 4 = 5$ while E has the score $1 + 2 = 3$. C still has the score 2. Again we select
the one with the smallest score, C, and proceed as before. We repeat this until
we reach I. Thus, the process is completed with the nodes having the following
scores H(4), G(6), F(6) and I(7).

Thus the shortest route from A to I has the score 7, and by tracing back
through adjacent nodes with the smallest score we get the path ABEHI.

In general terms, we can describe this process in the following algorithm.

> Initialise various identifiers that keep a record of our progress
> WHILE the number of nodes fixed is less than the number of nodes
>   Adjust the scores of the nodes reachable from the current node
>   Find the smallest score assigned to nodes that is greater than the
>   value of the current node
>   add one to the number of nodes fixed

This is illustrated in the procedure *shortroute*. The parameter *w*, of type *weights*,
is a square matrix of INTEGER which contains the weights corresponding to the
links that exist. If a direct link between two nodes does not exist, then the
weight is a large value. The *logicalmatrix m* contains the adjacency matrix for
the network. Finally, the parameter *score* returns the final scores for each of
the nodes in an array of type INTEGER.

```
PROCEDURE shortroute(w : weights; m : logicalmatrix;
 numberofnodes : INTEGER; VAR score : integervector);
VAR  i,j,term,smallestsofar : INTEGER;
     nodesfixed,temp,nextmin : INTEGER;
BEGIN

  score[1] :=0;
  term :=1;
  smallestsofar := 0;
  nodesfixed := 1;
  FOR i := 2 TO numberofnodes DO
       score[i] := 1000;
  WHILE nodesfixed <> numberofnodes DO
  BEGIN

 (*  adjust  the scores of the nodes that are reachable
     from the current node *)
    FOR j := 1 TO numberofnodes DO
      IF M[term,j] AND (score[j] > smallestsofar) THEN
         BEGIN
           temp := score[term] + w[term,j];
           IF temp < score[j] THEN score[j] := temp;
         END;

 (* find the the smallest score that is greater than the
    current one *)
    nextmin := 1000;
    FOR j := 1 TO numberofnodes DO
      IF (score[j] < nextmin) AND
                  (score[j] > smallestsofar) THEN
         BEGIN
           nextmin := score[j];
           term := j;
         END;

    smallestsofar := nextmin;
    nodesfixed := nodesfixed + 1;
  END;
END;
```

20.5 Exercises

20.1. Write procedures to carry out pre-order and post-order traversals of a
binary tree. Examine the effect of these, and the in-order traversal
described in section 20.3.1, on formulae such as

$$a*b+c/d*e*(a+b)$$

entered as characters into an ordered tree.

   A pre-order traversal is achieved by visiting each vertex, then its left
subtree and finally its right subtree. A post-order traversal visits the left
subtree, then the right subtree before visiting the vertex itself.

20.2. Write a procedure to determine the shortest route once the scores for each node have been assigned.

20.3. Write a procedure to determine the depth of a tree.

20.4. Write a procedure to enter the weights in a directed graph once the relationship has been entered.

20.5. Modify the procedure subopt given in section 20.2.1 so that the selected route is stored in

(a) an array,

(b) a linked list.

Appendix A: Standard Functions and Procedures

Pascal has a range of functions and procedures, that perform a variety of tasks, as a part of the language. The following is a summary of those described in this text.

Arithmetic Functions

The following functions are available to perform various arithmetic operations. Note carefully the remarks about the types of the arguments and the results.

SQR(x)   produces the square of the number x. If x is real the result is real, while if x is an integer the result is an integer.

ABS(x)   gives the value of $|x|$ – that is, the absolute value of x. Again ABS(x) has the same type as x.

SQRT(x)   produces the square root of x. (Clearly, x must have a positive value, else an error results.)

SIN(x)   produces the sine of the angle x radians.

COS(x)   produces the cosine of the angle x radians.

ARCTAN(x) gives the inverse tangent of x radians.

EXP(x)   gives the value of e raised to the power of x.

LN(x)   gives the logarithm of x to base e. (x must have a positive value, else an error results.)

Transfer Functions

ROUND(x) produces the integer result of rounding the real value x to the nearest integer.

TRUNC(x) gives the integer resulting from truncating x – that is, simply removing the decimal part.

ORD(x)   gives the ordinal number of the value of the scalar type x within the set defined by the type of x.

CHR(x)   gives the character corresponding to the value of the integer type x.

283

Boolean Functions

ODD(x)   returns the value TRUE if the integer parameter x is an odd number.

EOLN(filename) has the value true if we have reached the end of a line in the file with the stated filename.

EOF(filename)   has the value TRUE if we have reached the end of the stated file.

Further Functions

SUCC(x)   supplies the value after the current value of the scalar type x within the ordered set of values that the scalar type can take.

PRED(x)   supplies the value before the current value of the scalar type x.

Procedures

NEW(x)   creates a new address for the pointer x.

DISPOSE(x) disposes of the address of the pointer x.

Appendix B: Reserved and Pre-defined Words

The following words are reserved within the Pascal language and cannot be redefined.

AND	ARRAY	BEGIN	CASE	CONST	DIV	DO
DOWNTO	ELSE	END	FILE	FOR	FUNCTION	GOTO
IF	IN	LABEL	MOD	NOT	OF	OR
OTHERWISE	PACKED	PROCEDURE	PROGRAM	RECORD	REPEAT	SET
THEN	TO	TYPE	UNTIL	VAR	WHILE	WITH

The following words are pre-defined in Pascal. The user may redefine them, if desired, but in practice this is not to be recommended.

INTEGER	REAL	BOOLEAN	CHAR	TEXT
INPUT	OUTPUT			
FALSE	TRUE			
MAXINT	(this takes the value of the largest integer available on the implementation of Pascal being used)			

The following names of procedures are also reserved, many of which are described in the text.

DISPOSE GET* NEW PACK* PAGE* PUT* READ READLN
RESET REWRITE UNPACK* WRITE WRITELN

Those marked with an asterisk are not covered in this text.

References

Box, G. E. P. and Müller, M. E. (1958). 'A note on the generation of random normal deviates', *Ann. Math. Stat.*, **29**, 610–611.

Clarke, G. M. and Cooke, D. (1985). *A Basic Course in Statistics*, 2nd edn, Arnold, London.

Dew, P. M. and James, K. R. (1983). *Introduction to Numerical Computation in Pascal*, Macmillan, London.

Jensen, K. and Wirth, N. (1978). *Pascal User Manual and Report*, 2nd edn, Springer-Verlag, Berlin.

Kendall, M. G. and Babbington-Smith, B. (1939). *Tables of Random Sampling Numbers*, Tracts for Computers XXIV, Cambridge University Press.

Maron, M. J. (1987). *Numerical Analysis – A Practical Approach*, Macmillan Inc., New York.

Morgan, B. J. T. (1984). *Elements of Simulation*, Chapman Hall, London.

Phillips, G. M. and Taylor, P. J. (1973). *Theory and Applications of Numerical Analysis*, Academic Press, London.

Press, W. H., Flannery, B. P., Teukolsky, S. A. and Vetterling, W. T. (1986). *Numerical Recipes*, Cambridge University Press.

Ripley, B. D. (1983). 'Computer generation of random variables – a tutorial', *Inst. Stat. Rev.*, **51**, 303–319.

Skvarcius, R. and Robinson, W. B. (1986). *Discrete Mathematics with Computer Science Applications*, Benjamin Cummings, California.

Tippett, L. H. C. (1927). *Random Sampling Numbers*, Tracts for Computers XV, Cambridge University Press.

Wilson, R. J. (1985). *Introduction to Graph Theory*, Longman, Marlow.

Index